Current Topics in Microbiology 161 and Immunology

Picornaviruses

Edited by V. R. Racaniello

With 12 Figures

Springer-Verlag
Berlin Heidelberg NewYork
London Paris Tokyo HongKong

Vincent R. Racaniello, Ph.D.

Columbia University
College of Physicians and Surgeons,
701 W. 168th Street,
New York, NY 10032,
USA

ISBN 3-540-52429-0 Springer-Verlag Berlin Heidelberg New York
ISBN 0-387-52429-0 Springer-Verlag New York Berlin Heidelberg

© Springer-Verlag Berlin Heidelberg 1990
Library of Congress Catalog Card Number 15-12910
Printed in Germany

Typesetting: Thomson Press (India) Ltd., New Delhi
Offsetprinting: Saladruck, Berlin; Bookbinding: B. Helm Berlin
2123/3020-543210–Printed on acid-free paper.

Preface

At this writing the decade of the 1980s is rapidly coming to a close, and it is an appropriate time to review the picornavirus field. During the past decade there has been a remarkable reemergence of interest in picornaviruses and a virtual explosion of experimentation. The renaissance of picornaviruses can be attributed to several developments near the beginning of the 1980s. In 1981 the nucleotide sequence of the first picornavirus genome, that of poliovirus, was determined, providing a genetic map that would be the basis for a number of experimental questions regarding gene function and expression (Kitamura et al., *Nature* 291: 547; Racaniello and Baltimore, *Proc Natl Acad Sci USA* 78: 4887). In the same year it was reported that a cloned cDNA copy of the poliovirus genome is infectious when transfected into cultured mammalian cells (Racaniello and Baltimore, *Science* 214: 916, 1981). This discovery, which enables construction of poliovirus mutants and recombinants, has since been used for the study of many picornaviruses. Furthermore, the availability of cloned copies of viral genomes permits manipulation of gene products apart from infected cells. Third, the use of hybridoma technology to generate anti-picornavirus neutralizing monoclonal antibodies permitted mapping of antigenic sites (for example, Evans et al., *Nature* 304: 459, 1983). Finally, at mid-decade the three-dimensional structures of poliovirus (Hogle et al., *Science* 229: 1358, 1985) and rhinovirus (Rossmann et al., *Nature* 317: 145, 1985) were solved. The structures provide a physical description of antigenic sites on the virion and serve as the framework for experiments on virion assembly and disassembly.

These discoveries are the basis for much of the work described in this volume. Included are six chapters which cover most aspects of the replication of picornaviruses. Viral infection begins with binding of virus to a cell receptor, which is the topic of the first chapter by Racaniello ("Cell Receptors for Picornaviruses"). Receptors for poliovirus and rhinovirus were recently identified, and the molecular clones encoding these cell surface proteins will be potent reagents for future studies. Once picornavirus RNA has entered the cell it is translated in a unique manner: ribosomes bind directly to an internal sequence within the 5'-noncoding region. This process has been well studied in poliovirus and is the subject of the chapter by Sonenberg ("Poliovirus Translation"). Picornaviral protein is produced as a precursor, and cleavage of this protein is required to produce functional gene products. The intricacies of protein processing, which is carried out by virus coded proteinases, is the subject of the third chapter, by Lawson and Semler ("Picornavirus Protein Processing: Enzymes, Substrates and Genetic Regulation"), giving a comprehensive analysis of these activities. Several of the picornavirus proteins are involved in the replication of the viral RNA genome. RNA synthesis has been best studied during the course of poliovirus infection. Despite research in this area for over 25 years, the mechanism of RNA synthesis in poliovirus-infected cells has not been elucidated. The current state of the field is discussed in the chapter by Richards and Ehrenfeld ("Poliovirus RNA Replication"). The product of the infectious cycle, infectious virions, are of icosahedral symmetry. Our understanding of how this structure relates to antigenicity has improved markedly in the last decade. The sequences in the capsid that give rise to antiviral antibodies is the subject of the fifth chapter by Minor ("The Antigenic Structure of Picornaviruses"). This volume would not be complete without a general consideration of the genetics of picornaviruses. From the early rudimentary yet important studies on poliovirus mutants, isolated by chemical mutagenesis, the field has progressed to the state where defined mutants are readily isolated by site-directed mutagenesis. This approach is of enormous power and enables dissection of gene function. The final chapter by Sarnow, Jacobson, and Najita reviews the state of poliovirus genetics that has progressed so rapidly as a result of the advent of infectious viral cDNA clones ("Poliovirus Genetics").

The work of the past 10 years demonstrates that picornaviruses hold considerable interest for many investigators. Indeed, it is likely that experimentation on this fascinating family of viruses will continue well into the next century.

Vincent R. Racaniello
November 1989, New York City   Columbia University

List of Contents

List of Contributors

(Their addresses can be found at the beginning of their respective contributions)

Cell Receptors for Picornaviruses*

V. R. RACANIELLO

1 Introduction

The surfaces of animal cells bear specific receptor sites for different viruses. These sites are the first part of the host cell that viruses encounter as they begin the infectious cycle. Receptor binding has been studied as long as other events in the viral life cycle, yet with the exception of the influenza virus receptor, sialic acid, these studies have not, until recently, resulted in the isolation and identification of viral receptors. Undoubtedly one reason for the slow progress in this field has been the difficulties encountered in purification of cell receptors for viruses. The application of recombinant DNA and monoclonal antibody technologies has recently resulted in the identification of receptors for five viruses. Two of these viruses are members of the picornavirus family, and therefore it is a suitable time to review our knowledge of picornavirus receptors, and consider the questions about these receptors that can now be addressed. This review will focus on the identification of cell receptors for several picornaviruses. Other aspects of the

———————
Department of Microbiology, College of Physicians and Surgeons of Columbia University, 701 W. 168th St., New York, NY 10032, USA
* Work carried out in the author's laboratory is supported by grants from the American Cancer Society and the National Institute of Allergy and Infectious Diseases

virus-receptor interaction, including studies of binding kinetics and the events that occur shortly after receptor binding, have been extensively reviewed and will not be covered here (CROWELL and LANDAU 1983; MARSH and HELENIUS 1989).

It is often concluded that cell receptors are the main determinant of viral host range and cell and tissue tropism. Cell receptors are necessary for viral infection, but they are by no means sufficient. Once a virus binds a cell receptor, it must either be taken into the cell or fuse with the external envelope; its genetic material must be transcribed, translated, and replicated (not necessarily in that order); proteins must be processed and new virions must be assembled, to result in a productive infection. Viral replication in different cells may be blocked at many steps in the viral life cycle other than receptor binding. It is therefore important to remember that cell receptors are not the sole determinants of viral host range and cell and tissue tropism.

Poliovirus was one of the first animal viruses to be intensely studied, and therefore its cell receptor has also received much attention. As we shall see, information on the properties of picornavirus receptors emerged slowly. For example, it was found that cell receptors for poliovirus were membrane proteins, and competition experiments showed that in general, different picornaviruses employed different receptors. However, it was not until 1989 that two picornavirus receptors—those for poliovirus and for the major group of rhinoviruses— were identified. In the next several years approaches similar to those used for identifying the receptors for poliovirus and the major group of rhinoviruses will be used to identify other picornavirus receptors.

The picornavirus family contains the polioviruses, coxsackieviruses, echoviruses, rhinoviruses, cardioviruses (encephalomyocarditis virus, mengovirus, Theiler's murine encephalomyocarditis virus), aphthoviruses (foot-and-mouth disease virus), and hepatitis A viruses. Despite their similar genome organization and virion structures, these viruses cause a wide variety of human and animal illnesses, from paralytic poliomyelitis to the common cold to foot-and-mouth disease. Picornaviruses are comprised of a single, positive strand of RNA contained within a protein shell of icosahedral symmetry that is assembled from four different viral polypeptides. The viral RNAs contain a single long open reading frame that is translated directly upon entering the cytoplasm into a polyprotein that is proteolytically processed to form functional viral polypeptides. A consideration or our recent understanding of other aspects of picornavirus replication may be found in other chapters of this volume.

2 Cell Receptors for Poliovirus

2.1 Identification of a Cell Receptor

The idea that initiation of poliovirus replication requires a cell surface receptor (PVR) took hold in the 1950s and probably originated from analogies made with well-characterized bacteriophage systems and with other animal viruses such as influenza virus. Early experiments showed that poliovirus attached to cultured monkey kidney cells and human HeLa cells at an exponential rate, and this attachment was electrostatic, salt-dependent, and temperature-independent (YOUNGNER 1955; FOGH 1955; BACHTOLD et al. 1957; HOLLAND and MCLAREN 1959). When HeLa cells were disrupted by freeze-thawing, the resulting cell debris could specifically bind poliovirus and neutralize its infectivity (HOLLAND and MCLAREN 1959). Only debris prepared from poliovirus-susceptible primate cells, and not from poliovirus-resistant mouse cells, could neutralize infectivity, and therefore it was suggested that the active virus-binding material in these subcellular preparations was a cellular receptor (HOLLAND and MCLAREN 1959).

After these initial studies, general information on the physical properties of poliovirus receptors emerged slowly. Poliovirus binding activity in HeLa cell debris was found to be present mainly in the microsome fraction (HOLLAND and MCLAREN 1961). The virus-binding activity was sensitive to heat, ether, chloroform, and trypsin, but not to periodate, lipase, or neuraminidase (HOLLAND and MCLAREN 1959, 1961). Based on these studies, it was suggested that the receptor was a lipoprotein (HOLLAND and MCLAREN 1959, 1961). However, in a later study it was shown that binding of poliovirus to its receptor is blocked by concanavalin A, leading to the suggestion that this receptor is a glycoprotein (LONBERG-HOLM 1975b).

Treatment of whole cells with proteases inactivates receptors for poliovirus, which can regenerate within 12–16h (ZAJAC and CROWELL 1965b). The regeneration of receptors is prevented by treatment of cells with inhibitors of protein synthesis (LEVITT and CROWELL 1967). After treatment of cells with proteases, no receptors are detected on internal structures (ZAJAC and CROWELL 1965a).

Early attempts to solubilize the PVR from cell membranes failed (HOLLAND and MCLAREN 1961; DESENA and MANDEL 1976). Subsequently, when a solid-phase assay was used to detect virus binding, solubilization of the poliovirus receptor from membranes was achieved with deoxycholate, suggesting that the poliovirus receptor is an integral membrane protein (KRAH and CROWELL 1982). However, no further characterization of the receptor polypeptide was reported.

A combination of virus competition experiments and differential inactivation of binding activity by heat, treatment with proteolytic enzymes, or low pH treatment have demonstrated that the three poliovirus serotypes share a common receptor which is different from the receptor for the six serotypes of coxsackie-virus B, the group A coxsackieviruses, echoviruses, and the human rhinoviruses (CROWELL 1963, 1966; ZAJAC and CROWELL 1965a, b; LONBERG-HOLM et al. 1976). Virus binding studies have revealed that HeLa cells contain approximately 3×10^3 poliovirus receptor sites (LONBERG-HOLM and PHILIPSON 1974), although it is not known how many receptor molecules comprise a cellular receptor site.

It is well known that poliovirus replicates only in cultured cells of human or monkey origin, and not in cells of other animal species. For example, poliovirus does not replicate in various cells lines derived from the rabbit, dog, cat, swine, calf, guinea pig, mouse, or chick (MCLAREN et al. 1959). To determine whether this host range was controlled by the presence or absence of cell receptors, virus attachment to a variety of cultured cells was studied. The results indicated that susceptible human and monkey cell lines absorbed poliovirus, while insusceptible nonprimate cell lines such as those derived from the hamster, rabbit, dog, cat, swine, calf, guinea pig, mouse, and chick, did not absorb virus (KAPLAN 1955; MCLAREN et al. 1959). However, purified poliovirus RNA can initiate one round of replication in all cultured mammalian cells tested, including all the nonsusceptible nonprimate cells described above, demonstrating the absence of postreceptor blocks to poliovirus replication (DESOMER et al. 1959; HOLLAND et al. 1959a). These studies indicated that host range of poliovirus, at least in cultured cells, appears to be largely controlled at the stages of virus entry, such as receptor binding, virus penetration, and uncoating. However, not all resistance to poliovirus infection can be explained by blocks to virus entry (for example, see OKADA et al. 1987; KAPLAN et al. 1989).

Somatic cell hybrids between human and mouse cells were isolated to map the gene encoding a poliovirus receptor. Three cell lines that were susceptible to poliovirus infection contained 11–17 human chromosomes, and all included chromosome 19. Poliovirus-resistant lines derived from these cells all lacked chromosome 19. It was therefore suggested that susceptibility to poliovirus infection, perhaps mediated by the cell receptor, was encoded by a gene on human chromosome 19 (MILLER et al. 1974).

Several groups have isolated monoclonal antireceptor antibodies for use in characterizing the cell receptor for poliovirus. These monoclonal antibodies prevent virus binding to susceptible cells, and also protect cells from virus infection, presumably by blocking a critical epitope on the receptor. Using Hep2C cells as inoculum, two monoclonal antibodies, called 280 and 281, were isolated that protect cells against infection with all three poliovirus serotypes, but not with a variety of other viruses, including rhinovirus 1b, coxsackievirus A9 and B5, and nine other enteroviruses (MINOR et al. 1984). A second antibody, D171, directed against an epitope on HeLa cells, also blocked infection with all three poliovirus serotypes, but not with echovirus 30, coxsackievirus B5, herpes simplex virus type 1, vesicular stomatitis virus, and adenovirus 5 (NOBIS et al. 1985). All three antibodies appear to recognize the same antigenic site (P. NOBIS, personal communication). D171 binds only to cells of human or monkey origin, and not to pig, dog, rabbit, rat, hamster, or mouse cells, thereby confirming earlier conclusions that these nonsusceptible cells lines lack receptors. A third antibody, AF3, directed against HeLa cells, blocks infection with poliovirus types 1 and 2, but has little effect against type 3, and does not inhibit infection with echovirus 6, coxsackievirus B1, B3, and A18, influenza virus, and adenovirus 2 (SHEPLEY et al. 1988). The reason for the differential inhibition among the poliovirus serotypes is not known. AF3 detects a 100 kDa polypeptide in the membrane of HeLa cells in Western blot analysis. Recognition of proteins by antibodies 280, 281, and D171 has not been reported.

Since biochemical approaches had not been successful for isolating the cell receptor for poliovirus, a genetic approach was taken to identify genomic and cDNA sequences encoding the receptor (MENDELSOHN et al. 1986, 1989). The strategy used DNA-mediated transformation to transfer susceptibility to poliovirus infection from HeLa cells to nonsusceptible, receptor-negative mouse L cells. Mouse L cells were cotransformed with HeLa cell DNA and a selectable marker, and transformants were screened for susceptibility to poliovirus infection and for reactivity with antireceptor monoclonal antibody D171. Using this approach, mouse cell lines were isolated that were susceptible to infection with all three poliovirus serotypes. Next, genomic libraries were prepared from the L cell transformants, and the library was screened with the human Alu repeat sequence blur-8. The resulting genomic clones encoded the cell receptor for poliovirus, since they hybridized with DNAs in human but not mouse cells, and hybridized with a 3.3 kb RNA in HeLa cells and in the L cell transformants but not in untransformed L cells.

The genomic clones were used as hybridization probes to isolate cDNA clones from a HeLa cell library. Two long cDNA clones were isolated which, when transformed into L cells under the control of an SV40 promoter, directed the synthesis of functional poliovirus receptors as assayed by susceptibility to poliovirus infection. The cDNA clones therefore encode functional cell receptors for poliovirus.

The polypeptide encoded by the cDNA clones is a typical transmembrane protein, and contains an N-terminal hydrophobic signal sequence, an ecto-domain, a transmembrane domain and a cytoplasmic tail (Fig. 1). The two cDNA clones encode slightly different proteins of predicted molecular weight 43 and 45 kDa, which differ in the length of the cytoplasmic tail.

Examination of the predicted amino acid sequence reveals that the poliovirus receptor is a new member of the immunoglobulin superfamily. The predicted polypeptide can be folded into three domains, each of which is formed by disulfide bonding (Fig. 1). Other members of this superfamily have roles in cell-cell recognition and adhesion (WILLIAMS and BARCLAY 1988). Perhaps the natural cellular function of the poliovirus receptor involves such activities.

Although the predicted molecular weight of a receptor polypeptide is 43 or 45 kDa, the constitution of the receptor complex in the cell membrane is not known. To address this question, we have prepared polyclonal rabbit antiserum directed against a fusion polypeptide, synthesized in *E. coli*, of the *trpE* gene product and the PVR. These antibodies react with several polypeptides in HeLa cells, and efforts are currently directed at determining the relationship of these polypeptides to virus binding (M. FREISTADT and V. RACANIELLO, unpublished results).

Some information on the constitution of the virus binding site has been obtained through expression of the PVR in insect cells (G. KAPLAN and V.R. RACANIELLO, unpublished results). When the PVR is expressed in insect cells using a baculovirus vector, the receptor is expressed on the cell surface, and poliovirus is able to bind to the insect cells. Analysis with polyclonal antireceptor

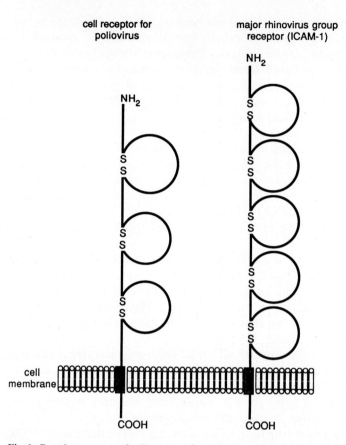

Fig. 1. Putative structure of cell receptors for poliovirus and the major rhinovirus group (ICAM-1). The ectodomain, with immunoglobulin-like domains formed by disulfide bonds, the transmembrane domain, and the cytoplasmic tails are shown. The actual constitution of the receptor sites is not known

antiserum indicates that the insect cells express a 67 kDa glycoprotein. A polypeptide of similar molecular weight is also observed in HeLa cells. These results suggest that the poliovirus binding site contains only the 67 kDa polypeptide, although it is not known how many polypeptides constitute the binding site.

Binding of poliovirus to insect cells expressing the PVR does not lead to productive infection. Although the block to poliovirus replication in insect cells is not known, it is possible that virus entry does not occur, which would suggest than an additional polypeptide is required for this step.

The predicted PVR polypeptide contains eight potential N-linked glycosylation sites. At least some of these are probably used since the receptor polypeptide produced in insect cells is 67 kDa, and treatment of these cells with tunicamycin results in a polypeptide of 35 kDa. Sialic acid may be a component of the poliovirus receptor, as revealed by the finding that poliovirus binding to cells is blocked by wheat germ lectin (TOMASSINI et al. 1989b).

An interesting question is the relationship between the 100 kDA polypeptide identified by monoclonal antibody AF3, and the PVR encoded in the HeLa cell cDNA clones. Our polyclonal antireceptor antisera do not detect polypeptides of 100 kDa in HeLa cells (unpublished results). AF3 might recognize a protein encoded by one of the larger receptor-related RNAs that are detected by Northern blot analysis (MENDELSOHN et al. 1989). Alternatively, AF3 might recognize a different protein that is associated with the receptor site, but not encoded by the PVR cDNA clones, possibly explaining the differential blocking of poliovirus infection by AF3.

2.2 Receptors and Host Range

As discussed above, restriction of poliovirus replication to primate cell cultures appears to be controlled by the cell receptor. In animals, susceptibility to poliovirus infection is largely confined to primates. To determine whether this restriction is also controlled at the stages of viral entry, viral RNA was inoculated intracerebrally into rabbits, chicks, guinea pigs, and hamsters. In all cases, virus was produced in the absence of disease (HOLLAND et al. 1959b). These authors concluded that, in the animals tested, the block to poliovirus replication is due to lack of receptors. However, these studies could not rule out blocks at the stages of virus penetration and uncoating, which would be bypassed by transfection with viral RNA.

Some poliovirus strains have been adapted to replicate in nonprimate hosts such as the mouse (ARMSTRONG 1939), chick embryo (ROCA-GARCIA et al. 1952), and suckling hamster (MOYER et al. 1952). Poliovirus variants capable of growing in mice have arisen spontaneously during propagation of viruses in cultured cells (LI and SCHAEFFER 1953; LI et al. 1955). Recent studies on the mouse-adapted P2/Lansing strain provide insight into why some poliovirus strains cannot replicate in mice. P2/Lansing was originally isolated from a fatal case of human poliomyelitis and subsequently adapted to mice (ARMSTRONG 1939). After intracerebral inoculation into mice, P2/Lansing induces fatal poliomyelitis. In contrast, inoculation of mice with P1/Mahoney fails to result in disease. Recently it was shown that the ability of P2/Lansing to infect mice can be transferred to the P1/Mahoney strain by exchange of the 8 amino acid sequence in VP1 known as neutralization antigenic site 1 (N-Ag1) (MURRAY et al. 1988; MARTIN et al. 1988). This result suggests that N-Ag1 of P2/Lansing is involved in receptor binding or virus entry in the mouse.

An important question is therefore whether the expanded host range of P2/Lansing is due to acquisition of a new receptor specificity. Unfortunately it is not possible to demonstrate binding of P2/Lansing to mouse brain homogenates (HOLLAND 1961), and therefore it cannot be determined whether receptors for P1/Mahoney are absent in the mouse. Our approach to this problem is to molecularly clone the P2/Lansing receptor cDNA from mice and determine whether it can bind P2/Lansing but not P1/Mahoney. We have isolated cDNA clones from mouse brain that encode a murine homolog of the human PVR that might represent the P2/Lansing receptor (M. MORRISON and V. RACANIELLO, unpublished results). The encoded murine protein is similar to the human PVR in that it has three Ig-like domains, a transmembrane sequence, and a cytoplasmic tail. The Ig-like domains of the human and mouse polypeptides have approximately 70% amino acid identity. Experiments are under way to determine whether the mouse cDNA encodes a functional receptor that can bind P2/Lansing but not P1/Mahoney.

2.3 Receptors and Tissue Tropism

Within the infected primate, poliovirus replication is limited to specific cells and tissues. For example, in humans poliovirus replicates in the oropharyngeal mucosa, in Peyer's patches of the gut, and in motor neurons within the central nervous system, and significant replication in other tissues has not been observed (BODIAN 1959). Poliovirus fails to replicate when inoculated directly into monkey kidney or testicular tissue (KAPLAN 1955; EVANS et al. 1954; LEDINKO et al. 1951). A simple explanation for this restricted tissue tropism is that it is determined by receptor distribution.

An early approach to this question was to examine the ability of poliovirus to bind to homogenates of tissues. This assay depends on the loss of infectivity associated with receptor binding at 37° C. In one study, it was found that virus bound to homogenates from susceptible organs of humans and monkeys (brain, spinal cord, intestine), while insusceptible tissues (heart, lung, kidney, skeletal muscle) did not bind virus (HOLLAND 1961). Virus binding was abolished by heating at 60° C, or by treatment with trypsin or 8 M urea. However, occasional low levels of virus binding to kidney, liver and lung were observed (HOLLAND 1961). Based on these studies, it was concluded that poliovirus tissue tropism is determined by the presence or absence of receptors. However, a separate study reported significant virus binding to many monkey and human tissues, including those that do not support poliovirus replication (KUNIN and JORDAN 1961). The reasons for the different results of the two studies are not known, but clearly a different approach to the problem of tissue tropism is required.

The availability of receptor cDNAs enables further examination of the relationship of tissue tropism and expression of cell receptors for poliovirus. When RNA from a variety of human tissues was examined by northern blot analysis, a 3.3 kb receptor RNA was detected in all tissues, including those which are sites of poliovirus replication (cerebellar cortex, motor cortex, spinal cord, ileum) and tissues in which poliovirus does not replicate (frontal cortex, kidney). Thus susceptibility to poliovirus infection does not appear to be governed solely by expression of receptor RNA in tissues.

It is curious that receptor RNAs were detected in tissues such as kidney that are not sites of poliovirus replication. Perhaps the receptor RNA is not translated into protein in these tissues, or perhaps the encoded protein cannot bind poliovirus because of differences in amino acid sequence or posttranslational modification. The 100 kDa protein detected by AF3 in HeLa cells is also present in human spinal cord, brain stem, and cortex, but not in kidney (SHEPLEY et al. 1988). AF3 appears to recognize a carbohydrate epitope (M. SHEPLEY, personal communication), and absence of this epitope in the kidney suggests that glycosylation is important for susceptibility to poliovirus infection. Perhaps expression of the PVR is limited to specific cells within the kidney that may not have access to infectious virus, or which contain other blocks to viral replication. We are currently employing our polyclonal antireceptor antibody to examine receptor expression in tissues by Western blot analysis, and in specific cell types by immunohistochemistry.

It was once believed that attenuation of neurovirulence of poliovirus vaccine strains was due to altered receptor specificity of the vaccine strains. This conclusion was based on the reduced ability of the vaccine strains to bind to central nervous system homogenates (SABIN 1957; KUNIN 1962; HOLLAND 1961). Subsequently another study found no difference in the binding of attenuated and neurovirulent virus strains (HARTER and CHOPPIN 1965). The location in the capsid of an attenuating mutation in the P2/Sabin strain is consistent with alteration of receptor binding (R. REN, E. MOSS, and V. RACANIELLO, unpublished results). However, the effects of the mutation are likely to be subtle and not detectable by assays of virus binding to tissues homogenates.

Although poliovirus replicates in few human tissues, with some exceptions it can replicate in nearly all cultured human cells. For example, injection of poliovirus into monkey kidney does not result in virus multiplication, and fresh kidney does not bind poliovirus, yet cultured monkey kidney cells are susceptible to poliovirus infection (KAPLAN 1955). The same observations were made using human amnion and cultured human amnion cells (HOLLAND 1961). Perhaps cell culture results in unmasking, modification, or synthesis of receptors to permit virus infection. The mechanism by which cultured cells acquire susceptibility to poliovirus should be clarified studying receptor expression with DNA and antibody probes.

3 Cell Receptors for Rhinovirus

Human rhinoviruses (HRVs), the agents of the common cold, infect only humans and chimpanzees. In humans, viral replication is limited to the upper respiratory tract. Most rhinovirus strains replicate only in human or chimpanzee cell lines.

Cell receptors for rhinovirus were first detected by adsorbing virus to cells HAFF et al. 1966; STOTT and HEATH 1970; THOMAS et al. 1970); adsorption was prevented by prior treatment of cells with trypsin (STOTT and HEATH 1970). The first clue that different rhinoviruses might employ different receptors came from the observation that, while HRV-2 and HRV-14 both adsorbed to WI-26 (human) cells, only HRV-2 adsorbed to African green monkey kidney cells (HAFF et al. 1966). This observation provided an explanation for the inability of HRV-14 to grow in AGMK cells. Subsequently competition studies demonstrated that HRV-2 and HRV-14 bind to different cell receptors (LONBERG-HOLM and KORANT 1972). Serotypes 3, 5, 15, 39, 41, and 51 were shown to employ the same receptor as HRV-14 (LONBERG-HOLM et al. 1976). More recently it was demonstrated that the vast majority of rhinovirus serotypes—approximately 78, including the prototype HRV-14—recognize one receptor, which has been termed the "major group" rhinovirus receptor (ABRAHAM and COLONNO 1984; COLONNO et al. 1986). Competition studies demonstrate that the major group receptor is also recognized by coxsackieviruses A13, A18, and A21 (LONBERG-HOLM et al. 1976). The remaining ten serotypes of HRV, including the prototypes HRV-1A and HRV-2, recognize what has been called the "minor group" receptor. The major and minor group HRV receptors are distinct from cell receptors used by poliovirus and CVB (LONBERG-HOLM et al. 1976).

For many years little information on the properties of rhinovirus receptors was available. Since virus binding was sensitive to trypsin (STOTT and HEATH 1970) and concanavalin A (LONBERG-HOLM 1975), the receptor was believed to be a cell surface glycoprotein. Progress on studying this receptor was probably hampered by its low abundance on cells. However, the development of monoclonal antibody technology enabled great advances in the study of HRV receptors. Using HeLa cells as immunogen, a monoclonal antibody was isolated that blocked attachment of rhinovirus to cells and protected cells against rhinovirus infection. This antibody, called 1A6, blocked infection only with rhinovirus serotypes of the major group was thus specific for the major group receptor. As would be predicted, 1A6 also blocked infection with coxsackie viruses A13, A18, and A21 (COLONNO et al. 1986).

Antibody 1A6 was used in binding assays to demonstrate that a wide variety of human cell lines, but not monkey, mouse, bovine, or goat cells, express the major HRV receptor (COLONNO et al. 1986). The major group receptor was also detected on chimpanzee liver cells, which are permissive for viral replication. Therefore, the presence of the major group receptor paralleled the known host range of the major group viruses. The same results were obtained when receptors were assayed by binding studies with labeled HRV.

Monoclonal antibody 1A6 was shown to bind to membrane extracts solubilized with sodium deoxycholate. Therefore, it was possible to isolate the receptor protein by gel filtration and monoclonal antibody affinity chromatography, using a solid phase radioimmunoassay to monitor purification. A polypeptide of apparent molecular weight 90 kDa on SDS-PAGE was eluted from the affinity column with diethylamine. When this protein was inoculated into rabbits, the resulting polyclonal antiserum blocked HRV-15 binding to HeLa cells, providing strong evidence that the 90 kDa polypeptide was the major group rhinovirus receptor, or at least a component of that receptor (TOMASSINI and COLONNO 1986). Further characterization of the 90 kDa polypeptide revealed that it was an acidic glycoprotein whose molecular mass was composed of 30% carbohydrate, including sialic acid, as revealed by digestion with various glycosidases (TOMASSINI et al. 1989b). Seven N-linked glycosylation sites were detected. However, neuraminidase treatment of cells reduced virus binding by only 35% and had no effect on binding of antibody 1A6. Therefore, it appears that sialic acid does not constitute the entire virus binding site.

Amino acid sequencing of CNBr and tryptic peptides derived from the purified 90 kDa receptor polypeptide enabled synthesis of oligonucleotide probes which were used to isolate cDNAs encoding the polypeptide (TOMASSINI et al. 1989a). A full length cDNA was constructed that directed the synthesis of functional HRV-15 and 1A6 binding sites in Vero cells after DNA transformation. This work proved that the 90 kDa polypeptide is the major group rhinovirus receptor.

Sequence analysis of the major group HRV receptor cDNA revealed that the encoded 55 kDa polypeptide is nearly identical to the previously described intercellular adhesion molecule-1 (ICAM-1; TOMASSINI et al. 1989a). The same conclusion was reached by two other groups. (GREVE et al. 1989; STAUNTON et al. 1989). One group isolated monoclonal antibodies that block attachment of the major group rhinoviruses to cells (GREVE et al. 1989). One monoclonal antibody was used to purify a 95 kDa polypeptide that bound rhinovirus; the amino acid sequence of this protein revealed its identity with ICAM-1. In addition, L cell transformants were obtained, using HeLa cell DNA as donor, that express functional rhinovirus binding sites. ICAM-1 oligonucleotide probes were used to obtain a full length cDNA from the transformants that directed the synthesis of rhinovirus binding sites after transformation into L cells. The gene encoding the ICAM-1 cDNA was mapped to human chromosome 19, the same chromosome that carries the gene for the poliovirus receptor. A third group had previously isolated full length cDNA clones encoding ICAM-1, based on its involvement in cell adhesion (STAUNTON et al. 1988). These workers subsequently demonstrated that transformation of an ICAM-1 cDNA clone into mouse cells results in expression of rhinovirus binding sites. Furthermore, purified ICAM-1 binds rhinovirus, and anti-ICAM-1 monoclonal antibodies prevent rhinovirus infection of HeLa cells.

ICAM-1 is a ligand for lymphocyte function-associated antigen 1 (LFA-1), a member of the integrin superfamily. The interaction between the two molecules is important for leukocyte adhesion to different cell types. Like the poliovirus receptor, ICAM-1 is a member of the immunoglobulin superfamily. However, ICAM-1 contains five predicted Ig-like domains, while the PVR contains three (Fig. 1).

It was previously known that synthesis of ICAM-1 is induced by soluble mediators of inflammation, such as interferon and interleukin (DUSTIN et al. 1986). As expected, treatment of HeLa cells with IFN-γ and TNF-α results in enhancement of HRV-15 binding (TOMASSINI et al. 1989a). It is therefore possible that the immune response to rhinovirus infection might induce receptor synthesis, thereby enhancing infection and spread of virus within the host. We must consider rhinovirus quite smart to have selected ICAM-1 for its cellular receptor.

Does the distribution of ICAM-1 in humans parallel rhinovirus tissue tropism? Studies on the expression of ICAM-1 indicate that it is found in a variety of human cell types, including vascular endothelial cells, thymic epithelial cells, other epithelial cells, fibroblasts, tissue macrophages, mitogen-stimulated T lymphocyte blasts, and germinal center dendritic cells in tonsils, lymph nodes, and Peyer's patches (DUSTIN et al. 1986). Indeed, the wide distribution of ICAM-1 was predicted by analyses with monoclonal antibody 1A6, which reacts with cells derived from a variety of human tissues (COLONNO et al. 1986). However, rhinovirus only replicates in epithelial cells of the upper respiratory tract, and therefore rhinovirus tropism cannot be explained by receptor distribution. Clearly other factors must regulate rhinovirus tissue tropism, such as the sensitivity of rhinovirus to low pH and high temperature.

It is curious that binding of rhinovirus to ICAM-1 expressed on mouse cells does not lead to productive infection (GREVE et al. 1989). Therefore, resistance of mouse cells to rhinovirus infection cannot be entirely attributed to the absence of viral receptors. The block to rhinovirus infection in ICAM-1 positive mouse cells has not been identified, but might involve any phase of viral replication after receptor binding.

Work on the rhinovirus minor group receptor is just beginning. In contrast to ICAM-1, minor group receptors are present on both human and mouse cells (COLONNO et al. 1986). Although HRV-2 will bind to mouse cells, viral replication does not occur. An HRV-2 variant selected for the ability to replicate in mouse cells has an altered 2C polypeptide, suggesting that the block to HRV-2 replication in these cells involves some aspect of viral RNA synthesis. (YIN and LOMAX 1985). Levels of the minor group receptor do not increase after treatment of cells with cytokines, as observed for ICAM-1 (TOMASSINI et al. 1989a). Binding of minor group rhinoviruses to cells is prevented by wheat germ lectin (TOMASSINI et al. 1989b), and to a lesser extent by concanavalin A (LONBERG-HOLM 1975). The minor group receptor will bind virus after solubilization of HeLa cells with detergents (MISCHAK et al. 1988). The solubilized receptor complex has been enriched by lectin and gel chromatography, and appears to have a molecular weight of 450 kDa. Identification of the receptor protein awaits molecular cloning of its cDNA and functional expression in cultured cells.

There are some clues about what part of rhinovirus attaches to ICAM-1. Examination of the three-dimensional structure of rhinovirus 14 reveals a prominent depression in the virion surface surrounding each fivefold axis of symmetry. This feature was called the "canyon" and it was suggested to be the site on the virion that interacts with the cell receptor (ROSSMANN et al. 1985). The canyon appears to be inaccessible to antibodies and therefore would not undergo change as a result of immune selection; sequence conservation would be an expected property of a receptor attachment site. To provide evidence in support of the canyon hypothesis, rhinovirus mutants were isolated that contain single amino acid changes at positions lining the canyon (COLONNO et al. 1988). Seven small plaque mutants were recovered that had lower binding affinities than wild type virus, suggesting that this region is involved in receptor binding. Further support for the canyon hypothesis comes from studies of a series of antiviral WIN compounds (Sterling-Winthrop) that block HRV-14 attachment to cells (HEINZ et al. 1989). These compounds insert into the virion in a hydrophobic pocket below the canyon floor and result in conformational changes in this area. Presumably the changes of the canyon result in a reduced ability of the virus to bind cell receptors. Eighty viral mutants resistant to these compounds were isolated, and the amino acid sequence associated with resistance was identified for each mutant. The results indicate that mutants resistant to high levels of the compound map at two sites lining the drug binding pocket, while mutants resistant to low levels contain changes at five sites in the region of the canyon altered by drug binding. One mechanism of drug resistance appears to involve compensation for the effect of the drug on virus attachment. Thus the mutations appear to lie in sites that are involved in virus attachment. Final identity of the virion attachment site will come when the three-dimensional structure of the virus-receptor complex is solved.

4 Cell Receptors for Coxsackievirus

Coxsackievirus infection of humans is associated with a wide variety of illnesses, including paralytic disease, meningitis, myocarditis, pleurodynia, exanthems, and enanthems. Thus coxsackieviruses display a broader tissue tropism than poliovirus or rhinovirus. Coxsackieviruses are classified in two groups A and B (CVA, CVB) according to the type of lesions produced after inoculation into suckling mice. Group A viruses produce diffuse myositis, while group B viruses produce focal areas of degeneration in the brain and muscle, and occasionally in the myocardium.

Coxsackievirus A9 adsorbs to a variety of susceptible primary human and rhesus monkey cell cultures, and infectivity is lost after incubation of virus with debris prepared from these cells (MCLAREN et al. 1960). Continuous human cell lines cannot adsorb coxsackievirus A9, but express receptors for coxsackieviruses B1, B3, and B5. Although the continuous lines are resistant to infection with coxsackie A viruses, the block can be bypassed by transfection with purified viral RNA, suggesting that receptors for the group A and B viruses are different (MCLAREN et al. 1960). Competition experiments confirm that the two virus groups use different receptors on HeLa cells (LONBERG-HOLM et al. 1976). Surprisingly, these studies also reveal that the CVB receptor is shared by adenovirus type 2. Other studies have demonstrated that CVB receptors are sensitive to chymotrypsin (ZAJAC and CROWELL 1965b), are present only on the cell surface (ZAJAC and CROWELL 1965a), and are regenerated after inactivation with proteolytic enzymes (ZAJAC and CROWELL 1965a; LEVITT and CROWELL 1967).

The receptor for group B coxsackieviruses has been solubilized from HeLa cell membranes, using sodium deoxycholate (KRAH and CROWELL 1982). To detect the receptor, a solid-phase assay was employed that measured binding of virus to plastic cell culture wells coated with solubilized membranes. This assay was used to further characterize the CVB3 receptor (KRAH and CROWELL 1985). The receptor complex has an apparent molecular weight of 275 kDa as determined by gel filtration. Virus binding activity is destroyed by treatment with glycosidases, suggesting that the receptor is a glycoprotein. This suggestion is supported by the observation that concanavalin A, and wheatgerm, lentil and pea lectins bind to the receptor complex. However, only concanavalin A, and pea and lentil lectins reduce virus attachment, suggesting that the sugar residues recognized by these lectins might be near the site that interacts with virus.

The ability to solubilize the CVB3 receptor was used as part of a receptor purification scheme. CVB3 virions were bound to HeLa cells, and the virus-receptor complex was solubilized with sodium deoxycholate and triton X-100 (MAPOLES et al. 1985). This complex, when purified from virions by centrifugation, contains the virion capsid proteins and one additional protein of 49.5 kDa called Rp-α. Rp-α purified from this complex can bind CVB3 and CVB1 but not poliovirus type 1. It was suggested that Rp-α is a member of the CVB receptor complex that is responsible for recognition and binding of the virion. Final proof of this hypothesis will require isolation of cDNA encoding Rp-α and demonstration that this cDNA directs expression of functional CVB binding sites. Once the CVB3 receptor is molecularly cloned, it will be interesting to determine whether it can also be recognized by Ad2, as suggested by virus competition studies (LONBERG-HOLM et al. 1976).

Monoclonal antibodies have also been useful for studying CV receptors. One group has reported the isolation of several monoclonal antibodies that block binding and infection of HeLa cells with CVB strains but not with PV1, CVA strains, or encephalomyocarditis virus (EMCV; CAMPBELL and CORDS 1983). These results confirmed the receptor specificities predicted by competition experiments discussed above. Another group has isolated anti-HeLa cell monoclonal antibodies that block attachment of CVB3 to cells (CROWELL et al. 1986). This monoclonal antibody protects cells from infection with CVB1, 3, 5, echovirus 6, and CVA21 but not from infection with four additional echovirus serotypes, four other CVA strains, HRV14, three poliovirus strains, three HRV strains, and EMCV. Some of these findings do not agree with the results of binding competition experiments. For example, competition studies show that CVA21 and HRV14 share the same receptor, while echovirus 6 does not share a receptor with CVB3. These differences remain unresolved.

An interesting observation has been made concerning receptor specificity of the coxsackie B viruses (REAGAN et al. 1984). These viruses do not replicate in human rhabdomyosarcoma (RD) cells because the cells do not bear CVB receptors. However, after blind passage in RD cells, CVB variants were obtained that could grow in these cells. These viral variants acquired the ability to agglutinate human erythrocytes, and formed small plaques on HeLa cells. In competition studies in HeLa cells, the CVB3-RD variant blocks attachment of CVB3, but CVB1 does not block receptors for CVB3-RD. These results suggest that the CVB3-RD variant recognizes a new receptor on RD cells, and that both this receptor and the normal CVB receptor are present on HeLa cells. Using HeLa cells as immunogen, monoclonal antibodies were obtained that could either block CVB3 infection of HeLa cells or CVB3-RD infection of RD cells. In addition, one monoclonal antibody, Rmc CVB3-CVB3RD, blocks infection of HeLa cells with CVB3 and infection of RD cells with CVB3-RD but has no effect on CVB3-RD infection of HeLa cells. This result is somewhat puzzling, because if the monoclonal antibody blocks each viral receptor in the respective cells, it would be expected to block infection of CVB3-RD in HeLa cells as well. An explanation of these observations might come from molecular cloning of the different CVB receptors on HeLa and RD cells.

5 Cell Receptors for Aphthovirus

Very little information is available about cell receptors for foot-and-mouth disease virus (FMDV). Early experiments examined the kinetics of FMDV adsorption to susceptible cells (THORNE and CARTWRIGHT 1961, BROWN et al. 1961, 1962; THORNE 1962; BAXT and BACHRACH 1980). Virus attaches to debris derived from susceptible cells, and heating the debris at 56° C, or treatment with lipid solvents or low pH (3.0) abolishes binding activity (THORNE and CARTWRIGHT 1961). The number of receptor sites on HHK-21 cells for several FMDV strains is estimated at $1-2.5 \times 10^4$ (BAXT and BACHRACH 1980). FMDV receptors reside in a plasma membrane fraction derived from BHK-21 cells, and virus does not bind to membranes prepared from trypsin-treated cells (BAXT and BACHRACH 1982). Competition experiments, employing radioactively labeled and unlabeled viruses, indicate that cells contain different receptors for the SAT, A, C, and O viral subtypes (BAXT and BACHRACH 1982; SEKIGUCHI et al. 1982).

Perhaps the best clues about the nature of FMDV receptors have come from analysis of the putative attachment site on the virion. Trypsin treatment of FMDV virions abolishes infectivity by preventing virus binding to cells (BROWN et al. 1963). The trypsin-treated virions have reduced density and reduced immunogenicity in guinea pigs (WILD and BROWN 1967). Electrophoretic analysis of the trypsin-treated virions indicates altered mobility of one of the capsid polypeptides (WILD et al. 1969). These data led to the suggestion that trypsin treatment removes a site on the virion responsible for attachment to a cell receptor (WILD et al. 1969). Since trypsin treatment also reduced the immunogenicity of the particles, it was suggested that the same part of the virion is also an immunizing antigen (WILD et al. 1969). A portion of VP1 between amino acids 135 and 160 was subsequently shown to constitute a major neutralization antigenic site, and several trypsin cleavage sites are located in this region (STROHMAIER et al. 1982; BITTLE et al. 1982). This part of VP1 contains the amino acid sequence Arg-Gly-Asp (RGD) at positions 145 to 147, which is conserved in the seven FMDV serotypes, with one exception, a RSGD in strain A1061. RGD is responsible for the binding of many extracellular ligands to cell surface receptors known as integrins (reviewed in RUOSLAHTI and PIERSCHBACHER 1987). For example, the extracellular glycoprotein fibronectin binds to a receptor via the RGD sequence, and RGD-containing peptides will attach to such receptors (PIERSCHBACHER and RUOSLAHTI 1984). Recently it was shown that RGD-containing peptides inhibit attachment of FMDV to BHK cells (Fox et al. 1989). Antibody directed against this region of VP1 blocks attachment of FMDV to BHK cells, and neutralizing monoclonal antibodies, which also block virus attachment, do not block virus binding in the presence of RGD-containing peptides. These result suggest that cell receptors for FMDV might be members of the integrin superfamily. It should be noted that trypsin treatment of FMDV also cleaves within the carboxy terminal region of VP1, in the area of residues 203–213, and this cleavage also results in inhibition of virus attachment (Fox et al. 1989). This region may therefore also be part of the receptor attachment site.

RGD-containing peptides block attachment of both A and O FMDV subtypes, although competition experiments had previously shown that these subtypes recognize two different receptors (Fox et al. 1989; BAXT and BACHRACH 1982; SEKIGUCHI et al. 1982). This result is not unexpected, because both fibronectin and vitronectin receptors are blocked by RGD-containing peptides, although the two proteins do not compete for the same receptor (PYTELA et al. 1985). Thus receptor specificity in FMDV is probably controlled by sequences in addition to the RGD, perhaps those at the carboxy terminus of VP1.

Since FMDV can infect many different cell types, DNA-mediated transformation probably cannot be used for cloning the FMDV receptor gene. It may be necessary to isolate monoclonal antibodies that inhibit FMDV attachment to cells, and use them to isolate the receptor protein. The receptor cDNAs could then be isolated using oligonucleotide probes derived from the receptor amino acid sequence, as was done for the major group rhinovirus receptor.

6 Cell Receptors for Other Picornaviruses

Nearly nothing is known about cell surface receptors for the many picornaviruses that have not been discussed here. The cardiovirus genus contains viruses such as mengovirus, encephalomyocarditis virus (EMCV), and Theiler's murine encephalomyocarditis virus (TMEV), all of which are pathogens of mice. Attachment of EMCV to human and mouse cells has been studied (MCCLINTOCK et al.

1980), but no other information on the receptor is available. An assay for TMEV receptors has recently been described in which virus is first bound to cells, followed by antiviral antibody and ^{125}I-labeled protein A (RUBIO and CUESTA 1988). Using this assay, specific binding of TMEV to BHK-21 cell receptors was demonstrated. Perhaps this assay will be used to provide information on the nature of TMEV receptors. Several picornaviruses, including echovirus 7 and 19, coxsackie B3, and EMCV, are able to bind to and agglutinate human erythrocytes. The receptor on erythrocytes for such viruses has been partially characterized (PHILIPSON et al. 1964), and in the case of EMCV is suggested to be glycophorin A (ALLAWAY et al. 1986). However, since attachment of viruses to erythrocytes does not lead to productive infection, it is not clear that study of these erythrocyte receptors will provide information about the receptors on susceptible cells.

7 Concluding Remarks

Now that cell receptors for two picornaviruses have been identified and their cDNAs molecularly cloned, a variety of interesting experiments are possible. One important goal is to provide a molecular description of the virus-receptor complex. To achieve this goal, it will first be necessary to identify the domain(s) of the PVR receptor and ICAM-1 that bind virus, followed by a more detailed analysis of the virus binding site by site-directed mutagenesis. This information can then be interpreted in the context of the three-dimensional structure of the receptor and the virus-receptor complex. The results of these studies would not only be scientifically fascinating, but might lead to the design of new antiviral agents.

It is possible that the steps that occur just after receptor binding, such as virus entry into the cell and uncoating, are in part mediated by the receptor protein. It will be of interest to determine whether receptor mutants can be isolated that display defects in mediating these activities. Perhaps entry and uncoating of picornaviruses are mediated by different polypeptides within the receptor complex.

Our understanding of the virus-receptor interaction would not be complete without a description of the part of the virion that interacts with the cell receptor. Some evidence has been presented which suggests that rhinovirus binds to ICAM-1 via the canyon. The structure of mengovirus reveals not a canyon, but rather "pits" surrounding the fivefold axis, and these have been proposed as cell attachment sites (LUO et al. 1987). It was a surprise, however, when the structure of FMDV revealed the absence of pits or canyons on the virion surface (ACHARYA et al. 1989). The RGD sequence within VP1, which appears to mediate binding of FMDV to cell receptors, is located in a polypeptide loop that forms a disordered protrusion on the virion surface, for which the structure could not be deduced.

Perhaps some picornaviruses bind to their receptors via surface loops, as in FMDV, while others bind to receptors via canyons or pits in the virion surface. Poliovirus may use a canyon or a loop to bind to cell receptors, depending on the animal host. The structure of poliovirus reveals a canyon like that of rhinovirus (HOGLE et al. 1985), although no evidence has been presented that this structure mediates binding to a cell receptor. The ability of P2/Lansing to infect mice can be conferred to another poliovirus strain by exchange of N-AgI, a polypeptide loop on the surface of the virion also known as the B-C loop (MURRAY et al. 1988; MARTIN et al. 1988). If infection of mice by P2/Lansing involves recognition of a new receptor by this strain, then it would appear that binding to this receptor is mediated by the B-C loop. Thus poliovirus host range variants that have acquired a second receptor specificity may recognize a different receptor in each host.

It is interesting that two members of the picornavirus family use receptors that are members of the immunoglobulin superfamily. Is this a coincidence, simply because so many surface proteins are members of this superfamily, or is the part of these viruses that binds the receptor particularly suited for attachment to members of the Ig superfamily? For example, might the domain structure of the Ig-like proteins fit nicely into the canyon of certain picornaviruses? Computer modeling of the C1 domain of ICAM-1, although very preliminary, suggests that C1 fits well into the rhinovirus canyon, and covers residues thought to be important for viral attachment (GIRANDA et al. 1989). It is clearly necessary to determine whether or not cell receptors for other picornaviruses are members of the Ig superfamily or the integrin superfamily, and whether virions bind to these receptors via canyons, pits, or protein loops.

In addition to receptors for rhinovirus and poliovirus, other receptors identified so far include that for HIV-1, known to be CD4, also an Ig superfamily member (MADDON et al. 1986). The receptor binding protein of HIV-1 is gp120; it will be interesting to determine whether the site on gp120 that binds receptor has a pit or canyon-like feature. Other receptors identified include the ecotropic Moloney murine leukemia virus receptor, a 67 kDa protein that may span the membrane seven times (ALBRITTON et al. 1989); complement receptor 2 (CR-2); the Epstein-Barr virus receptor (AHEARN et al. 1988; MOORE et al. 1987); and sialic acid, the influenza virus receptor.

Clearly the study of viral receptors is just beginning, and soon our knowledge of these structures and the ways in which they initiate the infectious cycle will approach what is known about other stages of viral replication. Although these studies will clarify yet another stage of the viral life cycle, receptors do not exist solely to be used by viruses. Identification of viral receptors in some cases means discovery of new cell proteins, and unravelling the function of these proteins will surely be an interesting task.

References

Abraham G, Colonno RJ (1984) Many rhinovirus serotypes share the same cellular receptor. J Virol 51: 340–345

Acharya R, Fry E, Stuart D, Fox G, Rowlands D, Brown F (989) The three-dimensional structure of foot-and-mouth disease virus at 2.9 Å resolution. Nature 337: 709–716

Ahearn J, Hayward S, Hickey J, Fearon D (1988) Epstein-Barr virus (EBV) infection of murine L cells expressing recombinant human EBV/C3d receptor. Proc Natl Acad Sci USA 85: 9307–9311

Albritton LA, Tseng L, Scadden D, Cunningham JM (1989) A putative murine ecotropic retrovirus receptor gene encodes a multiple membrane-spanning protein and confers susceptibility to virus infection. Cell 57: 659–666

Allaway GP, Pardoe IU, Tavakkol A, Burness ATH (1986) Encephalomyocarditis virus attachment. In: Crowell RL, Lonberg-Holm K (eds) Virus attachment and entry into cells. American Society for Microbiology, Washington DC

Armstrong C (1939) Cotton rats and white mice in poliomyelitis research. Public Health Rep 54: 1719–1721

Bachtold JG, Bubel HC, Gebhardt LP (1957) The primary interaction of poliomyelitis virus with host cells of tissue culture origin. Virology 4: 582–589

Baxt B, Bachrach HL (1980) Early interactions of foot and mouth disease virus with culture cells. Virology 104: 42–55

Baxt B, Bachrach HL (1982) The adsorption and degradation of foot-and-mouth disease virus by isolated BHK21 cell plasma membranes. Virology 116: 391–405

Bittle JL, Houghten RA, Alexander H, Shinnick TM, Rowlands D, Brown F (1982) Protection against foot-and-mouth disease by immunization with a chemically synthesized peptide predicted from the viral nucleotide sequence. Nature 298: 30–33

Bodian D (1959) Poliomyelitis: pathogenesis and histopathology. In Horsfall FL, Tamm I (eds) Viral and rickettsial infections of man. Lippincott, Philadelphia

Brown F, Cartwright B, Stewart DL (1961) Mechanism of infection of pig kidney cells by foot-and-mouth disease virus. Biophys Biochim Acta 47: 172–177

Brown F, Cartwright B, Stewart DL (1962) Further studies on the infection of pig-kidney cells by foot-and-mouth disease virus. Biochim Biophys Acta 55: 768–774

Brown F, Cartwright B, Stewart DL (1963) The effect of various inactivating agents on the viral ribonucleic acid infectivities of foot-and-mouth and on its attachment to susceptible cells. J Gen Microbiol 31: 179–186

Campbell BA, Cords CE (1983) Monoclonal antibodies that inhibit attachment of group B coxsackieviruses. J Virol 48: 561–564

Colonno RJ, Callahan PL, Long WL (1986) Isolation of a monoclonal antibody that blocks attachment of the major group of human rhinoviruses. J Virol 57: 7–12

Colonno R, Condra J, Mizutani S, Callahan P, Davies M-E, Murcko M (1988) Evidence for the direct involvement of the rhinovirus canyon in receptor binding. Proc Natl Acad Sci USA 85: 5449–5453

Crowell RL (1963) Specific viral interference in HeLa cell cultures chronically infected with coxsackie B5 virus. J Bacteriol 86: 517–526

Crowell RL (1966) Specific cell-surface alternation by enteroviruses as reflected by viral attachment interference. J Bacteriol 91: 198–204

Crowell RL, Landau BJ (1983) Receptors in the initiation of picornavirus infections. In: Fraenkel-Conrat H, Wagner RR (eds) Comprehensive virology: Virus-Host Interactions: Receptors persistence and neurological diseases, vol 18. Plenum, New York

Crowell RL, Field AK, Schleif WA, Long WL, Colonno RJ, Mapoles JE, Emini EA (1986) Monoclonal antibody that inhibits infection of HeLa and rhabdomyosarcoma cells by selected enteroviruses through receptor blockade. J Virol 57: 438–445

DeSena J, Mandel B (1976) Studies on the in vitro uncoating of poliovirus I. Characterization of the modifying factor and the modifying reaction. Virology 70: 470–483

DeSomer P, Prinzie A, Schonne E (1959) Infectivity of poliovirus RNA for embryonated eggs and unsusceptible cell lines. Nature 184: 652–653

Dustin ML, Rothlein R, Bhan AK, Dinarello CA, Springer TA (1986) Induction by IL-1 and interferon, tissue distribution, biochemistry, and function of a natural adherence molecule (ICAM-1). J Immunol 137: 245–254

Evans CA, Byatt PH, Chambers VC, Smith WM (1954) Growth of neurotropic viruses in extraneural tissues VI. Absence of in vivo multiplication of poliomyelitis virus, types I and II, after intratesticular inoculation of monkeys and other animals. J Immunol 72: 348–352

Fogh J (1955) Relation between multiplicity of exposure, adsorption, and cytopathogenic effect of poliomyelitis virus in monkey kidney tissue culture. Virology 1: 324–333

Fox G, Parry NR, Barnett PV, McGinn B, Rowlands DJ, Brown F (1989) The cell attachment site on foot-and-mouth disease virus includes the amino acid sequence RGD (arginine-glycine-aspartic acid). J Virol 70: 625–637

Giranda VL, Chapman M, Rossmann MG (1989) Modelling of the C1 domain of the intracellular adhesion molecule 1 (ICAM-1), the human rhinovirus major group receptor. Abstracts, 2nd International Symposium on positive strand RNA viruses, Vienna, Austria, p 97

Greve JM, Davis G, Meyer AM, Forte CP, Yost SC, Marlor CW, Kamarck MF, McClelland A (1989) The major human rhinovirus receptor is ICAM-1. Cell 56: 839–847

Haff RF, Wohlsen B, Force EE, Stewart RC (1966) Growth characteristics of two rhinovirus strains in WI-26 and monkey kidney cells. J Bacteriol 6: 2339–2342

Harter DH, Choppin PW (1965) Adsorption of attenuated and neurovirulent poliovirus strains to central nervous system tissues of primates. J Immunol 95: 730–736

Heinz BA, Shepard DA, Rueckert RR (1989) Drug-resistant mutants of human rhinoviruses map to capsid regions involved in attachment. In: Europic 89. (abstr no. G10)

Hogle JM, Chow M, Filman DJ (1985) Three-dimensional structure of poliovirus at 2.9Å resolution. Science 229: 1358–1365

Holland JJ (1961) Receptor affinities as major determinants of enterovirus tissue tropisms in humans. Virology 15: 312–326

Holland JJ, McLaren LC (1959) The mammalian cell-virus relationship II. Adsorption, reception, and eclipse by HeLa cells. J Exp Med 109: 487–504

Holland JJ, McLaren LC (1961) The location and nature of enterovirus receptors in susceptible cells. J Exp Med 114: 161–171

Holland JJ, McLaren JC, Syverton JT (1959a) The mammalian cell virus relationship III. Production of infectious poliovirus by non-primate cells exposed to poliovirus ribonucleic acid. Proc Soc Exp Biol Med 100: 843–845

Holland JJ, McLaren JC, Syverton JT (1959b) The mammalian cell virus relationship IV. Infection of naturally insusceptible cells with enterovirus ribonucleic acid. J Exp Med 110: 65–80

Kaplan AS (1955) Comparison of susceptible and resistant cells to infection with poliomyelitis virus. Ann NY Acad Sci 61: 830–839

Kaplan G, Levy A, Racaniello VR (1989) Isolation and characterization of HeLa cell lines blocked at different steps in the poliovirus life cycle. J Virol 63: 43–51

Krah DL, Crowell RL (1982) A solid-phase assay of solubilized HeLa cell membrane receptors for binding group B coxsackieviruses and polioviruses Virology 118: 148–156

Krah DL, Crowell RL (1985) Properties of the deoxycholate-solubilized HeLa cell plasma membrane receptor for binding group B coxsackieviruses. J Virol 53: 867–870

Kunin CM (1962) Virus-tissue union and the pathogenesis of enterovirus infections. J Immunol 8: 556–559

Kunin CM, Jordan WS (1961) In vitro adsorption of poliovirus by noncultured tissues. Effect of species, age and malignancy. Am J Hyg 73: 245–257

Ledinko N, Riordan JT, Melnick JL (1951) Differences in cellular pathogenicity of two immunologically related poliomyelitis viruses as revealed in tissue culture. Proc Soc Exp Biol Med 78: 83–88

Levitt NH, Crowell RL (1967) Comparative studies of the regeneration of HeLa cell receptors for poliovirus T1 and coxsackievirus B3. J Virol 1: 693–700

Li CP, Schaeffer M (1953) Adaptation of type 1 poliomyelitis virus to mice. Proc Soc Exp Biol Med 82: 477–481

Li CP, Schaeffer M, Nelson DB (1955) Experimentally produced variants of poliomyelitis virus combining in vivo and in vitro techniques. Ann NY Acad Sci 61: 902–910

Lonberg-Holm K (1975) The effects of concanavalin A on the early events of infection by rhinovirus type 2 and poliovirus type 2. J Gen Virol 28: 313–327

Lonberg-Holm K, Korant BD (1972) Early interaction of rhinoviruses with host cells. J Virol 9: 29–40

Lonberg-Holm K, Philipson L (1974) Early interaction between animal viruses and cells. Monogr Virol 9: 1–148

Lonberg-Holm K, Crowell RL, Philipson L (1976) Unrelated animal viruses share receptors. Nature 259: 679–681

Luo M, Vriend G, Kamer G, Minor I, Arnold E, Rossmann MG, Boege U, Scraba DG, Duke GM, Palmenberg AC (1987) The atomic structure of mengovirus at 3.0Å resolution. Science 235: 182–191

Maddon PJ, Dalgleish AG, McDougal JS, Clapham PR, Weiss RA, Axel R (1986) The T4 gene encodes the AIDS virus receptor and is expressed in the immune system and the brain. Cell 47: 333–348

Mapoles JE, Krah DL, Crowell RL (1985) Purification of a HeLa cell receptor protein for group B coxsackieviruses. J Virol 5: 560–566

Marsh M, Helenius A (1989) Virus entry into animal cells. Adv Virus Res 36: 107–151

Martin A, Wychowski C, Couderc T, Crainic R, Hogle J, Girard M (1988) Engineering a poliovirus type 2 antigenic site on a type 1 capsid results in a chimeric virus which is neurovirulent for mice. EMBO J 7: 2839–2847

McClintock PR, Billups LC, Notkins AL (1980) Receptors for encephalomyocarditis virus on murine and human cells. Virology 106: 261–272

McLaren LC, Holland JJ, Syverton JT (1959) The mammalian cell-virus relationship I. Attachment of poliovirus to cultivated cells of primate and non-primate origin. J Exp Med 109: 475–485

McLaren LC, Holland JJ, Syverton JT (1960) The mammalian cell-virus relationship V. Susceptibility and resistance of cells in vitro to infection by coxsackie A9 virus. J Exp Med 112: 581–594

Mendelsohn C, Johnson B. Lionetti KA, Nobis P, Wimmer E, Racaniello VR (1986) Transformation of a human poliovirus receptor gene into mouse cells. Proc Natl Acad Sci USA 83: 7845–7849

Mendelsohn C, Wimmer E, Racaniello VR (1989) Cellular receptor for poliovirus: molecular cloning, nucleotide sequence and expression of a new member of the immunoglobulin superfamily. Cell 56: 855–865

Miller, DA, Miller, OJ, Dev VG, Hashmi S, Tantravahi R, Medrano L, Green H (1974) Human chromosome 19 carries a poliovirus receptor gene. Cell 1: 167–173

Minor PD, Pipkin PA, Hockley D, Schild GC, Almond JW (1984) Monoclonal antibodies which block cellular receptors of poliovirus. Virus Res 1: 203–212

Mischak H, Neubauer C, Kuechler E, Blaas D (1988) Characteristics of the minor group receptor of human rhinoviruses. Virology 163: 19–25

Moore MD, Cooper NR, Tack BF, Nemerow GR (1987) Molecular cloning of the cDNA encoding the Epstein-Barr virus/C3d receptor (complement receptor type 2) of human B lymphocytes. Proc Natl Acad Sci USA 84: 9194–9198

Moyer AQ, Accorti C, Cox HR (1952) Poliomyelitis I. Propagation of the MEF1 strain of poliomyelitis virus in the suckling hamster. Proc Soc Exp Biol Med 81: 513–518

Murray MG, Bradley J, Yang XF, Wimmer E, Moss EG, Racaniello VR (1988) Poliovirus host range is determined by a short amino acid sequence in neutralization antigenic site I. Science 241: 213–215

Nobis P, Zibirre R, Meyer G, Kuhne J, Warnecke G, Koch G (1985) Production of a monoclonal antibody against an epitope on HeLa cells that is the functional poliovirus binding site. J Gen Virol 6: 2563–2569

Okada Y, Toda G, Oka H, Nomoto A, Yoshikura H (1987) Poliovirus infection of established human blood cell lines: relationship between the differentiation stage and susceptibility to cell killing. Virol 156: 238–245

Philipson L, Bengtsson S, Brishammar S, Svennerholm L, Zetterquist O (1964) Purification and chemical analysis of the erythrocyte receptor for hemagglutinating enteroviruses. Virology 2: 580–590

Pierschbacher MD, Ruoslahti E (1984) Cell attachment activity of fibronectin can be duplicated by small synthetic fragments of the molecule. Nature 309: 30–33

Pytela R, Pierschbacher MD, Ruoslahti E (1985) A 125/115kDa cell surface receptor specific for vitronectin interacts with the arginine-glycine-aspartic acid adhesion sequence derived from fibronectin. Proc Natl Acad Sci USA 82: 8057–8061

Reagan KJ, Goldberg B, Crowell RL (1984) Altered receptor specificity of coxsackievirus B3 after growth in rhabdomyosarcoma cells. J Virol 49: 635–640

Roca-Garcia M, Moyer AW, Cox HR (1952) Poliomyelitis II. Propagation of MEF1 strain of poliomyelitis virus in developing chick embryo by yolk sac inoculation. Proc Soc Exp Biol Med 81: 519–525

Rossmann MG, Arnold E, Erickson JW, Frankenberger EA, Griffith JP, Hecht H-J, Johnson JE, Kamer G (1985) Structure of a human common cold virus and functional relationship to other picornaviruses. Nature 317: 145–153

Rubio N, Cuesta A (1988) Receptors for Theiler's murine encephalomyelitis virus: characterization by using rabbit antiviral antiserum. J Virol 62: 4304–4306

Ruoslahti E, Pierschbacher MD (1987) New perspectives in cell adhesion: RGD and integrins. Science 238: 491–497

Sabin AB (1957) Properties of attenuated polioviruses and their behavior in human beings. In: TM Rivers (ed) Cellular biology, nucleic acids and viruses. New York Academy of Science, New York

Sekiguchi K, Franke AJ, Baxt B (1982) Competition for cellular receptor sites among selected aphthoviruses. Arch Virol 74: 53–64

Shepley MP, Sherry B, Weiner HL (1988) Monoclonal antibody identification of a 100-kDa membrane protein in HeLa cells and human spinal cord involved in poliovirus attachment. Proc Natl Acad Sci USA 85: 7743–7747

Staunton DE, Marlin SD, Stratowa C, Dustin ML, Springer TA (1988) Primary structure of ICAM-1 demonstrates interaction between members of the immunoglobulin and integrin supergene families. Cell 52: 925–933

Staunton DE, Merluzzi VJ, Rothlein R, Barton R, Marlin SD, Springer TA (1989) A cell adhesion molecule, ICAM-1, is the major surface receptor for rhinoviruses. Cell 56: 849–853

Stott EJ, Heath GF (1970) Factors affecting the growth of rhinovirus 2 in suspension cultures of L132 cells. J Gen Virol 6: 15–24

Strohmaier K, Franze R, Adam KH (1982) Location and characterization of the antigenic portion of the FMDV immunizing protein. J Gen Virol 59: 295–306

Thomas DC, Conant RM, Hamparian VV (1970) Replication of rhinovirus. Proc Soc Exp Biol Med 133: 62–65

Thorne HV (1962) Kinetics of cell infection and penetration by the virus of foot-and-mouth disease. J Bacteriol 84: 929–942

Thorne HV, Cartwright SF (1961) Reactions of the virus of foot-and-mouth disease with cells and cell debris. Virology 15: 245–257

Tomassini JE, Colonno RJ (1986) Isolation of a receptor protein involved in attachment of human rhinoviruses. J Virol 58: 290–295

Tomassini JE, Graham D, DeWitt CM, Lineberger DW, Rodkey JA, Colonno RJ (1989a) cDNA cloning reveals that the major group rhinovirus receptor on HeLa cells is intercellular adhesion molecule 1. Proc Natl Acad Sci USA 86: 4907–4911

Tomassini JE, Maxson TR, Colonno RJ (1989b) Biochemical characterization of a glycoprotein involved in rhinovirus attachment. J Biol Chem 264: 1656–1662

Wild TF, Brown F (1967) Nature of the inactivating action of trypsin on foot-and-mouth disease virus. J Gen Virol 1: 247–250

Wild TF, Burroughs JN, Brown F (1969) Surface structure of foot-and-mouth disease virus. J Gen Virol 4: 313–320

Williams AF, Barclay AN (1988) The immunoglobulin superfamily—domains for cell surface recognition. Annu Rev Immunol 6: 381–405

Yin FH, Lomax NB (1985) Host range mutants of human rhinovirus in which nonstructural proteins are altered. J Virol 48: 410–418

Youngner JS (1955) Virus adsorption and plaque formation in monolayer cultures of trypsin-dispersed monkey kidney. J Immunol 76: 288–292

Zajac I, Crowell R (1965a) Location and regeneration of enterovirus receptors of HeLa cells. J Bacteriol 89: 1097–1100

Zajac I, Crowell RL (1965b) Effect of enzymes on the interaction of enteroviruses with living HeLa cells. J Bacteriol 89: 574–582

Poliovirus Translation*

N. SONENBERG

1 Introduction

Considerable effort has gone into investigations of translation of picornavirus and particularly of poliovirus. Picornavirus mRNAs have unique structural properties that make them ideal for studying the mechanism and control of translation initiation in eukaryotes. Accordingly, the studies of poliovirus and encephalomyocarditis virus (EMCV) mRNAs and their translation have yielded novel insights into the mechanism of eukaryotic translation. In addition, the ability of poliovirus to induce a precipitous and dramatic reduction of cellular protein synthesis stems from its unique translational features.

The poliovirus replication cycle commences with the binding of the virus to a specific receptor followed by internalization. In the cell the genomic RNA is uncoated and is then translated into a polyprotein that is processed in the native

Department of Biochemistry and McGill Cancer Center, McGill University, Montreal, Quebec, Canada
* The work cited from the author's laboratory was supported by the Medical Research Council of Canada

form by two virus-encoded proteinases into the different functional viral polypeptides (see Chap. 3, this volume). Translation in the presence of amino acid analogues results in the synthesis of a single polypeptide with a molecular weight of 247 000. The genomic RNA is positive stranded with ∼ 7500 nucleotides (different serotypes are slightly different in length), and contains a long 5' noncoding region of ∼ 740 nucleotides, and a much shorter 3' noncoding region of ∼ 70 nucleotides (see RACANIELLO and BALTIMORE 1981; KITAMURA et al. 1981) for poliovirus type 1, Mahoney strain. Sequences of the other poliovirus strains have been also determined.

Like most cellular and viral mRNA, poliovirus RNA contains a 3' poly A tail of heterogeneous length (YOGO and WIMMER 1972). However, unlike all cellular and most viral mRNAs the poliovirus genomic RNA does not possess a cap structure (m^{7}GpppN, see below), but instead has a small polypeptide termed VPg covalently linked to its 5' end (FLANAGAN et al. 1977; LEE et al. 1977). In polysomal RNA VPg is removed resulting in a 5' pUp terminated RNA (HEWLETT et al. 1976; NOMOTO et al. 1976).

The long 5' noncoding region of poliovirus RNA is laden with six to eight AUG codons of which three are conserved among the three different serotypes (TOYODA et al. 1984). These AUGs are not likely to serve as initiation codons, because mutation of six out of the seven upstream AUGs in type 2 (Lansing strain) have no phenotypic effect. (As discussed below the seventh AUG is also not likely to serve as an initiation codon.) In addition, no translation products from the upstream small ORFs have been identified. It is significant, however, that the major portion (up to position ∼ 600) of the 5' UTR (untranslated region) of polioviruses is conserved, and thus must serve an important function. One of these functions is the initiation of protein synthesis.

Elucidation of the mechanisms of translation of poliovirus RNA and the shut-off of host protein synthesis was greatly facilitated by the increased knowledge of the translation initiation process in eukaryotes. Therefore, before describing the "unique" translational properties of poliovirus I will outline the pathway of initiation of translation of cellular mRNAs.

2 Initiation of Translation in Eukaryotes

Most of the understanding of translation initiation in eukaryotes is based on biochemical identification and characterization of initiation factor activities from rabbit reticulocyte lysate (for review see MOLDAVE 1985; PAIN 1986). The initiation pathway is shown in Fig. 1. Of special importance for the discussion of poliovirus translation and the mechanism of shut-off of host protein synthesis are two steps: (a) the binding of ribosomes to the mRNA; (b) the recycling of eIF-2 after its release from ribosomes as an [eIF-2·GDP] complex.

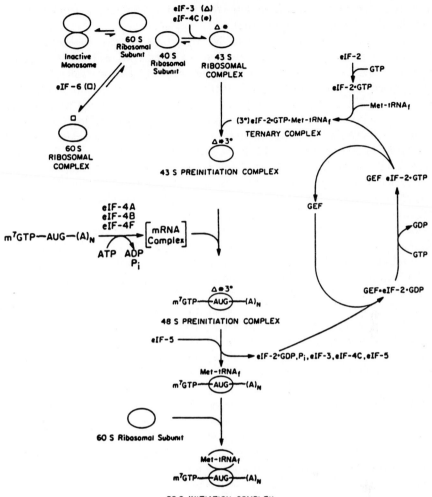

Fig. 1. 80S initiation complex formation in eukaryotes. The figure was modified from the model published by MERRICK et al. (1987) with permission from the authors. For further details see reviews by MOLDAVE (1985), PAIN (1986), and MERRICK et al. (1987)

2.1 Ribosome Binding to mRNA

This complex step in the initiation pathway is most probably the rate limiting step in translation initiation (JAGUS et al. 1981). Binding of ribosomes to mRNA requires *cis*-acting elements on the mRNA, three different initiation factors, and is dependent on ATP hydrolysis. Several elements in eukaryotic mRNAs play an important role in ribosome binding: these include the 5′ terminal cap structure, primary and secondary structure of the 5′ noncoding region, and possibly the 3′ noncoding region and the polyA tail.

The structure m⁷GpppN (where N is any nucleotide), termed the cap structure, is present at the 5′ terminus of all eukaryotic mRNAs (excluding organellar mRNAs; SHATKIN 1976). Most but not all eukaryotic viral mRNAs are also capped; the exceptions include picornaviruses, caliciviruses, and several plant viral RNAs. Numerous studies in vitro have demonstrated that the cap structure functions to facilitate ribosome binding to mRNA (for reviews see BANERJEE 1980; SHATKIN 1985). In addition, the cap structure was demonstrated to enhance nuclear processes such as pre-mRNA splicing (KONARSKA et al. 1984; EDERY and SONENBERG 1985) and 3′ end processing (GEORGIEV et al. 1984). The cap structure also stabilizes mRNA in the cytoplasm (FURUICHI et al. 1977) and pre-mRNAs in the nucleus (GREEN et al. 1983) against 5′ exonucleolytic degradation.

Several proteins that bind to the mRNA mediate cap functioning during translation initiation. A 24 kDa polypeptide termed eIF-4E (or CBPI) contains the binding site for the cap structure (SONENBERG et al. 1978). This protein exists in the cytoplasm as a singular protein or as a component in a three-subunit complex termed eIF-4F (also, cap-binding protein complex or CBPII; TAHARA et al. 1981; GRIFO et al. 1983; EDERY et al. 1983); eIF-4E in both forms can be cross-linked specifically to the mRNA 5′ structure, albeit the cross-linking is about tenfold more efficient when eIF-4E is part of eIF-4F (LEE et al. 1985a). In addition to eIF-4E, the CBP complex eIF-4F contains a 50 kDa polypeptide that is very similar to the previously characterized initiation factor eIF-4A (molecular weight ~ 50 000; GRIFO et al. 1983; EDERY et al. 1983). The third polypeptide in eIF-4F has an apparent molecular weight of 220 000 and was termed p220. As will be discussed below, proteolysis of this subunit is the apparent major cause of the inactivation of eIF-4F, bringing about the inability of cellular mRNA to translate in poliovirus-infected cells. eIF-4F and eIF-4A are both required for maximum translation in an in vitro reconstituted translation system (GRIFO et al. 1983). Thus, eIF-4A is required in a singular form and as part of eIF-4F for efficient translation. In contrast to eIF-4F, its small subunit eIF-4E was not active in the in vitro reconstituted translation system (GRIFO et al. 1983). This and other results suggest that eIF-4E functions in the cap recognition process as part of the three subunit complex, eIF-4F. There is, however, a report suggesting that eIF-4E functions as a singular entity in binding to the mRNA cap structure (HIREMATH et al. 1989).

eIF-4E was discovered by cross-linking of crude ribosomal high salt wash fractions to the mRNA 5′ cap structure in the absence of ATP (SONENBERG et al. 1978). Under these conditions, it was the only polypeptide that cross-linked specifically to the cap structure. However, when cross-linking was performed in the presence of ATP, two additional polypeptides eIF-4A and eIF-4B cross-linked to the cap structure in the presence of ATP (SONENBERG 1981; SONENBERG et al. 1981; GRIFO et al. 1982; EDERY et al. 1983). This led to the proposal that subsequent to the binding of eIF-4F to the cap structure a process of melting of

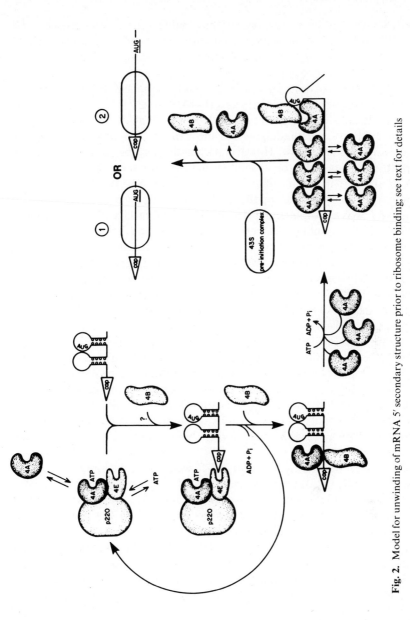

Fig. 2. Model for unwinding of mRNA 5' secondary structure prior to ribosome binding; see text for details

the 5' mRNA secondary structure occurs, thus enabling the interaction of eIF-4A and eIF-4B with the cap structure (SONENBERG et al. 1981). The necessity to unwind mRNA secondary structure for efficient translation is likely to be the major raison d'être of the cap structure.

Indeed eIF-4F functions as a helicase. The component in eIF-4F that is responsible for the helicase activity is eIF-4A, but eIF-4A on its own also possesses helicase activity (RAY et al. 1985; ROZEN et al. 1990). This feature is important for the understanding of the mechanism of poliovirus translation initiation as described below. However, depending on the assay used eIF-4B appears to stimulate or to be absolutely required for the helicase activity of eIF-4A or eIF-4F (RAY et al. 1985; ROZEN et al. 1990). Several observations are consistent with eIF-4A possessing the catalytic site for a helicase activity: (a) eIF-4A singularly or as a component of eIF-4F binds ATP specifically, as determined by UV induced cross-linking. ATP cross-linking to eIF-4A in the eIF-4F complex is, however, 60 times more efficient than free eIF-4A (SARKAR et al. 1985). (b) eIF-4A contains an ATP-binding consensus sequence and is part of a larger gene family (D-E-A-D family) whose proteins contain the sequence $AX_4^SG_4^SGKT$ (LINDER et al. 1989; ROZEN et al. 1989). Mutation of the lysine residue in this sequence abrogates ATP binding (ROZEN et al. 1989). (c) eIF-4A exhibits ATPase activity that is single-stranded RNA dependent (ABRAMSON et al. 1987). (d) eIF-4A functions as a helicase. It was shown to increase RNAase sensitivity of reovirus mRNA in an ATP-dependent and eIF-4B-stimulated manner (RAY et al. 1985). Furthermore, eIF-4A unwound DNA-RNA duplexes (LAWSON et al. 1989) or RNA-RNA duplexes (ROZEN et al. 1990). It is noteworthy that eIF-4B was absolutely required for unwinding of RNA-RNA duplexes with either eIF-4A or eIF-4F. (e) eIF-4F depleted of its eIF-4A component had no helicase activity (ROZEN et al. 1990) and was inactive in the RNAase sensitivity assay (RAY et al. 1985).

The model for unwinding of the mRNA 5' secondary structure by the eIF-4A subunit of eIF-4F in conjunction with eIF-4B is schematically illustrated in Fig. 2. It is not entirely clear whether eIF-4F and p220 are released from the mRNA, before mRNA unwinding, as depicted in the model. It is also postulated that free eIF-4A maintains the mRNA in the unwound single-stranded configuration by binding to the denatured region of the mRNA in a reaction that requires ATP hydrolysis (ABRAMSON et al. 1987). Ribosomes can bind to the mRNA only following secondary structure unwinding. However, it is not known where on the mRNA is the initial ribosomal entry site. This is indicated in Fig. 2 by two different possibilities.

2.2 eIF-2 Recycling

The reader is referred to two small reviews for more detailed description of this step in the regulation of translation initiation (SAFER 1983; PROUD 1986). Protein synthesis initiation factor eIF-2 is a multisubunit complex, comprising three nonidentical subunits designated α, β and γ. eIF-2 forms a ternary complex with GTP and methionyl-tRNA that is transferred to the 40S ribosomal subunit. eIF-2 is released from the ribosome as a complex with GDP prior to 60S subunit joining (see Fig. 1). The binding constant of GDP to eIF-2 is ~ 100-fold greater than that for GTP. Consequently, in order to recycle eIF-2 for a new round of initiation, GDP has to be exchanged for GTP. An exchange factor which contains multiple subunits has been identified and purified and was termed GEF (guanine nucleotide exchange factor, also eIF-2B, RF, eRF or SP; see reviews by SAFER 1983 and PROUD 1986). Phosphorylation of the α-subunit of eIF-2 by two different kinases, HCR (heme control repressor) and dsI (double-stranded RNA-activated kinase) results in the entrapment of GEF and consequently a block in eIF-2 recycling.

3 Cap-Independent Translation of Poliovirus RNA

Upon the discovery that poliovirus mRNA is naturally uncapped it became clear that its translation must proceed by a cap-independent mechanism. Indeed, in vitro experiments showed that translation of poliovirus RNA was insensitive to inhibition by cap-analogues, under conditions where translation of most capped mRNAs is strongly inhibited. In addition, poliovirus RNA translates in extracts from poliovirus-infected cells in which the cap-binding protein complex, eIF-4F is inactivated (ROSE et al. 1978). It has been suspected for some time that translation initiation occurs by direct ribosome binding to the 5' noncoding region of poliovirus mRNA (e.g., PEREZ-BERCOFF 1982). However, direct evidence has been obtained only recently. There are several lines of evidence that an internal sequence of the poliovirus 5' UTR is necessary and sufficient for translation initiation. First, sequences within the mRNA 5' noncoding region can confer cap-independent translation to a heterologous gene (PELLETIER et al. 1988a; TRONO et al. 1988b; PELLETIER and SONENBERG 1989). Mutational deletion analysis of poliovirus type 2 revealed that sequences between nucleotides 320 and 631 are required for cap-independent translation in extracts prepared from poliovirus-infected cells (PELLETIER et al. 1988a). These sequences were also found to be important for the translation of poliovirus type 1 in vivo in poliovirus infected cells and in vitro in cell extracts (TRONO et al. 1988b). Second, sequences between nucleotides 567 to 627 of poliovirus type 1 (Mahoney strain) were required for efficient translation in rabbit reticulocyte lysate supplemented with a HeLa extract (BIENKOWSKA-SZEWCZYK and EHRENFELD 1988). Third, the finding that an extended region of the 5' noncoding is required for cap-independent translation is supported by biochemical and genetic analysis of defined mutations in the 5' noncoding region (KUGE and NOMOTO 1987; TRONO et al. 1988a). Small deletions or insertions over a wide region (nucleotides 224, 270, and 392) in the 5' noncoding region yielded virus that made very little protein and did not inhibit host cell translation, but synthesized a significant amount of viral RNA—a phenotype consistent with a defect in translation. Based on this conclusion the region was termed P for protein synthesis (TRONO et al. 1988a). Several other mutations in region P were lethal, suggesting that this region is critical for translation. The notion that the P region extends hundreds of nucleotides is also supported by the results of KUGE and NOMOTO (1987), who generated a four-nucleotide insertion mutant at nucleotide 220 of the Sabin 1 virus. This mutant virus had similar properties to those described above by TRONO et al. (1988a). Partial revertants of the mutants were isolated and found to contain second-site mutations in nucleotides 186 and 524 or 186 and 480, suggesting that the functional length of region P might extend from nucleotide 184 to 524. Finally, a mutation of an A base at position 588 of poliovirus type 2 (Lansing strain) caused a small-plaque phenotype and reduced translational efficiency in a HeLa cell-free extract in vitro (PELLETIER et al. 1988c).

Direct evidence for internal binding of ribosomes to the poliovirus 5′ UTR was provided by PELLETIER and SONENBERG (1988, 1989), who used a bicistronic mRNA in which the poliovirus 5′ UTR was inserted into the intercistronic region. In vivo experiments showed that the second cistron could be translated under conditions in which the first cistron was not translated, i.e., in poliovirus-infected cells or cells grown in hypertonic medium. Thus, translation of the second cistron was separate and independent of that of the first. Control experiments showed that the independent translation of the second cistron did not result from fragmentation of the bicistronic mRNA (PELLETIER and SONENBERG 1988). In the absence of the intercistronic poliovirus 5′ UTR, translation of the second cistron was dependent on translation of the first cistron. Translation in extracts prepared from HeLa cells substantiated the conclusions drawn from the in vivo experiments (PELLETIER and SONENBERG 1989). In addition, translation of the second cistron in the TK/P2CAT mRNA, in contrast to the first cistron, was as predicted not sensitive to cap analogue inhibition in vitro (PELLETIER and SONENBERG 1989). The region that was required and sufficient for internal ribosome binding was termed RLP for ribosome landing pad.

Internal binding of ribosomes to the poliovirus 5′ UTR is also consistent with the finding that mutation of six out of the seven upstream AUGs in the 5′ UTR had no effect on translation in vitro or on virus replication in vivo (PELLETIER et al. 1988c). According to the ribosome scanning model (KOZAK 1983) upstream AUGs have inhibitory effects on translation and their removal is expected to increase translational efficiency if ribosomes access the initiator AUG by scanning.

From the current data it is not clear how the ribosome reaches the poliovirus initiator AUG at position 745. There are several possible alternative scenarios: (a) The ribosome binds to a specific sequence in the RLP and then linearly scans the RNA to encounter the initiator AUG. (b) The ribosome binds to a specific sequence in the RLP and then is translocated directly to the initiator AUG.

(c) The ribosome binds directly to the initiator AUG, in a manner that is dependent on the upstream RLP sequence. Several results suggest, but do not prove, that ribosomes bind upstream of the initiator AUG in the RLP region and then scan the RNA to encounter the initiator AUG. Insertion of RNA containing stable secondary structure between the RLP and the initiator AUG (at nucleotide 631) inhibited translation, whereas the insertion of the same sequence upstream of the RLP had no effect on translation (PELLETIER and SONENBERG 1988). Moreover, KUGE et al. (1989) found that insertion of an AUG codon between the RLP and the initiator AUG (in the nonconserved ∼ 100 nucleotide sequence), yielded a mutant virus exhibiting a small plaque phenotype. Wild type revertants contained mutations in the inserted AUG codon or deletion thereof. It is of interest that the 100-nucleotide sequence just upstream of the initiator AUG, in addition to being the least conserved sequence among polioviruses, does not contain any AUGs. Thus, insertion of an AUG or secondary structure (PELLETIER and SONENBERG 1988) in this region might interfere with the movement or translocation of the ribosome from its binding site to the initiator AUG.

The identification of a large sequence in the poliovirus 5′ UTR that participates in internal ribosome binding strongly argues for the involvement of secondary and tertiary structures in this process. Several models for the secondary structure of the 5′ UTR have been described (RIVERA et al. 1988; PILIPENKO et al. 1989a; SKINNER et al. 1989). These models were generated by computer-predicted minimal energy folding, comparative analysis of the secondary structure of polioviruses, coxsackie viruses and rhinoviruses, and biochemical probing with RNases and chemicals in solution. Although differences exist between the three secondary structure models, there are also striking similarities, particularly between the models of PILIPENKO et al. (1989a) and SKINNER et al. (1989). In the latter models, the region between nucleotides 240 and 620 contain three domains of secondary structure, termed I, II, and III, starting from the 5′ end (PILIPENKO et al. 1989a). There are strong suggestions that the secondary structure plays an important functional role. This structure is conserved among polioviruses, coxsackie B viruses, and rhinoviruses, and, most importantly, contain compensatory base-pair changes in which the base pairs are changed but the secondary structure is conserved (PILIPENKO et al. 1989b; SKINNER et al. 1989). In addition, attenuated viruses reverted to a virulent phenotype by restoring wild type secondary structure using compensatory mutations (SKINNER et al. 1989).

It is conceivable that the secondary and tertiary structure motifs in the RLP region are recognized by proteins that facilitate ribosome internal binding. One approach to identify such proteins is the gel-electrophoresis mobility shift assay. For the mobility shift assay, MEEROVITCH et al. (1989) used a small RNA fragment encompassing nucleotides 559 to 624. This RNA contains an adenosine at position 588 whose mutation reduced translational efficiency in vitro and displayed a small plaque phenotype (PELLETIER et al. 1988c). This fragment formed a specific complex with a component in a HeLa cell extract. A substitution of A_{588} to a uridine, which affects the template activity of the RNA, also reduced the extent of complex formation. In agreement with these results, DEL ANGEL et al. (1989) demonstrated specific binding of cytoplasmic factors with a sequence from nucleotides 510–629. In addition, they determined by RNase footprinting that protein binding occurred to a conserved stem-loop structure containing nucleotides 550 to 629. Complex formation was also observed with a second sequence (nucleotides 97–182) within the 5′ UTR (DEL ANGEL et al. 1989). The polypeptide that is complexed with the 559–624 RNA fragment was identified by cross-linking as a cellular protein having a molecular weight of 52 000 (MEEROVITCH et al. 1989). This protein (termed p52) does not seem to be one of the characterized translation initiation or elongation factors, and hence is probably a hitherto unidentified translation initiation factor. However, DEL ANGEL et al. (1989) found that eIF-2 is part of the complexes formed with sequences from nucleotides 97–182 and 510–629. The reason for the discrepancy between these results is not clear at present. It is of interest that the levels of p52 in reticulocyte lysates and wheat-germ extracts, in which poliovirus RNA is translated inefficiently, are much lower than in HeLa extracts in which translation is very efficient, suggesting that the limitation of p52 in the former translation systems is responsible for the inefficient translation (MEEROVITCH et al. 1989). This conclusion is strongly supported by recent findings that purified p52 stimulated preferentially the translation of poliovirus 5′ UTR containing mRNA in reticulocyte lysate (K. MEEROVITCH, unpublished results). It is also likely that p52 is the HeLa factor that corrects the aberrant translation pattern of poliovirus RNA is reticulocyte lysate as discussed below (SVITKIN et al. 1988).

3.1 Molecular Mechanism of Poliovirus Cap-Independent Translation

A model for poliovirus RNA translation initiation is shown in Fig. 3. The RLP region is first recognized by one (p52) or more cellular factors that assist in the subsequent binding of eIF-4A and eIF-4B. It is very likely that the secondary and tertiary structure of the RLP region are major determinants in the initial recognition by cellular factors including p52. Following binding of eIF-4A and eIF-4B, they catalyze the unwinding of the RNA secondary structure to create the binding site for the 40S ribosome. The salient feature of this model is that unwinding of RNA secondary structure initiates at an internal site on the RNA, in contrast to the generally accepted model for cellular capped mRNAs, where unwinding initiates near the cap structure (SONENBERG 1988). Indeed, it was recently shown that eIF-4A, in combination with eIF-4B, can unwind a partial duplex RNA that is double stranded at its ends and contains a single-stranded region in the middle, in a bidirectional manner (ROZEN et al. 1990).

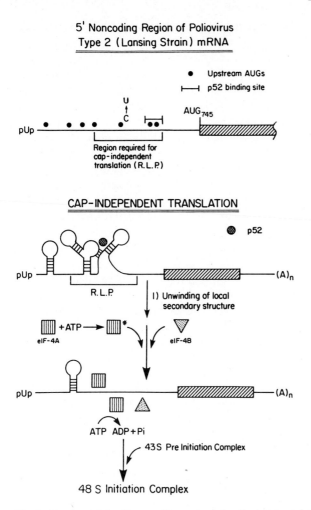

Fig. 3. Diagram of the 5′ noncoding region of poliovirus type 2 (Lansing strain) and model for internal ribosome binding. The model proposes that a cellular factor p52 is involved in the recognition of a specific sequence and/or structure within the RLP region followed by binding of eIF-4A and eIF-4B and consequent melting of the secondary structure to allow ribosome binding

3.2 Cap-Independent Translation of Other Picornaviruses

The 5′ structures of all picornavirus genomes have common features with that of poliovirus in that they do not possess a 5′ cap structure and contain long 5′ UTRs (700–1200 nucleotides) laden with multiple AUGs. Their translational strategy is therefore expected to resemble that of poliovirus. Indeed, translation initiation on EMCV RNA was demonstrated to occur by direct binding of ribosomes to an internal sequence in the 5′ noncoding region (JANG et al. 1988, 1989). Early indications of such a mechanism were reported by SHIH et al. (1987), who showed that hybridization of cDNA fragments to sequences between nucleotides 1 and 338 of the 5' UTR had no effect on translation of EMCV RNA. In contrast, hybridization of cDNAs corresponding to sequences between nucleotides 450 and 834 strongly inhibited translation. Subsequently, JANG et al. (1988, 1989) demonstrated both in vitro and in vivo that ribosomes can initiate internally on a bicistronic mRNA, in which the intercistronic region contained nucleotides 260 to 848 of the EMCV 5' UTR. This region was termed IRES for internal ribosome entry site. Primer extension analysis showed that the results were not due to nucleolytic degradation of the bicistronic mRNA (JANG et al. 1989). As for poliovirus, the EMCV IRES is comprised of hundreds of nucleotides, suggesting the existence of a superstructure that plays an important function in this process. An extensive secondary structure has been determined for the EMCV 5' UTR (PILIPENKO et al. 1989b), which is different from that described for poliovirus 5' UTR. Some translational properties of EMC viral RNA are different from poliovirus, particularly the ability of EMC viral RNA to translate efficiently in a reticulocyte lysate (SHIH et al. 1978; BROWN and EHRENFELD 1979; EMMERT and PHILLIPS 1986; DORNER et al. 1984). Moreover, following binding to the EMCV IRES, ribosomes are probably transferred directly to the initiator AUG, unlike the scanning process occurring on poliovirus RNA (R. JACKSON, personal communication). These results imply that the requirements of *trans*-acting factors for ribosome internal binding are different for EMCV and poliovirus.

Direct internal binding of ribosomes to eukaryotic mRNAs is not limited to naturally uncapped mRNAs such as poliovirus and EMCV, but was found also with other viral mRNAs. In most of these cases internal binding of ribosomes occurred to regions that are contained within translated ORFs, thus synthesizing overlapping proteins (HERMAN 1986; HASSIN et al. 1986; CURRAN and KOLAKOFSKY 1989; CHANG et al. 1989). In addition, there are other viral mRNAs whose translation is largely independent of the presence of the cap-binding protein complex (eIF-4F) including the major late promoter mRNAs of adenovirus (BABLANIAN and RUSSEL 1974; DOLPH et al. 1988; CASTRILLO and CARRASCO 1988), and alfalfa mosaic virus-4 RNA (GEHRKE et al. 1983). It is not clear, however, that ribosomes bind internally to these mRNAs.

In light of the finding that internal ribosome binding occurs in uninfected cells, it is conceivable that cellular mRNAs also use this mechanism of translation. It was speculated that cellular mRNAs with long 5' UTRs might be candidates for such a mechanism (PELLETIER and SONENBERG 1988). More recently, it was reported that translation of the glucose-regulated protein 78 is increased in poliovirus-infected cells when cap-dependent translation of cellular mRNAs is inhibited (SARNOW 1988). Also, heat-shock proteins are more resistant than the bulk of cellular mRNAs to inhibition after poliovirus infection (MUNOZ et al. 1984). Although it is clear that these mRNAs have a very reduced requirement for eIF-4F, it remains to be seen if ribosomes can bind internally within the 5' UTRs of these mRNAs.

4 Cell-Specific Translation of Poliovirus RNA

Translational efficiency of poliovirus RNA varies considerably among different cell extracts. Translation of poliovirus was initially shown to be less efficient than EMCV RNA in a reticulocyte lysate (SHIH et al. 1978). This result has engendered the view that poliovirus is in general a feeble translator (see for example, DORNER et al. 1984. However, this notion is erroneous since poliovirus RNA translates efficiently in extracts prepared from HeLa cells (PELLETIER et al. 1988b). More importantly, in vivo when EMCV infected HeLa cells are superinfected with poliovirus, poliovirus translation is not reduced (DETJEN et al. 1981; ALONSO and CARRASCO 1981), in spite of the fact that EMCV mRNA is believed to outcompete cellular mRNAs. Also, cardioviruses and poliovirus can replicate simultaneously (McCORMICK and PENMAN 1968; SHIRMAN et al. 1973). Therefore, both in vivo and in vitro, poliovirus RNA can translate efficiently. However, in several in vitro translation systems including rabbit reticulocyte lysate and wheat-germ extract (SHIH et al. 1978; PELLETIER et al. 1988b), and in vivo in *Xenopus* oocytes (PELLETIER et al. 1988b), translation is inefficient. The low translational efficiency of poliovirus RNA in certain translation systems was attributed to sequences in the 5′ UTR, since deletion of the 5′ UTR dramatically increased translational efficiency in rabbit reticulocyte lysate, and wheat-germ and *Xenopus* oocytes (NICKLIN et al. 1987; PELLETIER et al. 1988b). Deletion analysis mapped the major inhibitory sequence between nucleotides 70 and 381 (PELLETIER et al. 1988b).

How can the cell-specific translational restriction of poliovirus RNA be explained? It is possible that rabbit reticulocyte lysate, and wheat-germ or *Xenopus* oocytes are limiting in a factor that promotes internal binding of ribosomes to the 5′ UTR in a cap-independent fashion. Consequently, translation in these systems would be facilitated by a 5′ end mediated process, an inefficient process due to the inhibitory elements between nucleotides 70 and 380 (PELLETIER et al. 1988b). This region can assume a stable secondary structure as predicted by computer modelling and confirmed by chemical and RNase mapping (RIVERA et al. 1988; PILIPENKO et al. 1989a; SKINNER et al. 1989). Since secondary structure in the 5′ UTR of several mRNAs has been shown to inhibit cap-dependent translation in vitro and in vivo (PELLETIER and SONENBERG 1985b), it is conceivable that the secondary structure in the 5′ UTR of poliovirus interdicts translation. In HeLa cell extract (and also in L cells), however, ribosomes would bypass the translational barrier by binding internally to the RLP of poliovirus mRNA (PELLETIER and SONENBERG 1988).

The cell-specific translation probably has bearing on earlier observations that a significant proportion of translation of poliovirus mRNA in a rabbit reticulocyte lysate, but not in a HeLa extract, initiates in the P3 region located in the 3′ one-third of the RNA (BROWN and EHRENFELD 1979; DORNER et al. 1984; PHILLIPS and EMMERT 1986). In addition, electron microscopy studies showed ribosome binding to the mRNA P3 region in rabbit reticulocyte lysate (McCLAIN et al. 1981). This abnormal initiation could be reduced by the addition of components from HeLa cell extracts (BROWN and EHRENFELD 1979; DORNER et al. 1984; PHILLIPS and EMMERT 1986; SVITKIN et al. 1988). PHILLIPS and EMMERT (1986) presented evidence that reticulocyte lysate is limiting in a factor that promotes the utilization of the 5' proximal initiator AUG. In addition, several rabbit reticulocyte initiation factors (eIF-4A, -4B, -4F and 2) did not correct the aberrant initiation in reticulocyte lysate. These findings were confirmed and extended by PELLETIER et al. (1988b), who showed that ribosomal high salt wash preparations from HeLa cells, but not from rabbit reticulocytes, stimulated translation of a hybrid mRNA containing the poliovirus 5' UTR fused to the CAT ORF. SVITKIN et al. (1988) have partially purified an activity from Krebs-2 cells that altered the selection of translation sites on poliovirus RNA to favor the 5' proximal initiator AUG. This factor, termed initiation correcting factor (ICF), copurified with the eIF-2/eIF-2B complex. Consequently, it was suggested that ICF is equivalent to eIF-2/eIF-2B complex. This is not consistent with the finding that a rabbit reticulocyte ribosomal salt wash and purified eIF-2/eIF-2B had no ICF activity (SVITKIN et al. 1988; R. JACKSON, personal communication). However, the possibility, although unlikely, cannot be ruled out that rabbit reticulocyte eIF-2 or eIF-2B is modified differently than the corresponding HeLa factors. It is also unlikely that the ICF is eIF-4A, as suggested by the results of DANIELS-MCQUEEN et al. (1983), who showed that poliovirus RNA translation in rabbit reticulocyte lysate is enhanced by the addition of eIF-4A. Initiation factor-4A is a very abundant factor in rabbit reticulocyte lysate and addition of initiation factors or purified eIF-4A did not relieve the translational inhibition imparted by the 5' proximal sequence of the poliovirus mRNA (PELLETIER et al. 1988b). A likely candidate for ICF is HeLa p52— the HeLa protein that can bind to a specific sequence between nucleotides 559–624 in the poliovirus 5' UTR (MEEROVITCH et al. 1989), since it is limiting in reticulocyte lysate and the purified protein can stimulate translation in the reticulocyte lysate (K. MEEROVITCH, unpublished observations).

It would be important to determine whether cell-specific translation occurs in vivo. Viral replication in several established human blood cell lines was cell lineage and differentiation stage dependent (OKADA et al. 1987). It was not established whether viral replication was controlled at the translational lavel, but in light of the in vitro translation results this is an interesting possibility. The cell-specific translational restriction of poliovirus replication could also explain in part tissue tropism of poliovirus (HOLLAND 1961). Although it is believed that poliovirus tissue tropism is determined largely by the distribution of cell-specific virus receptors (HOLLAND 1961), it is possible that additional restrictions are imposed by the translational machinery in certain tissues. Thus, the cell-specific differential translation of poliovirus might have relevance to the limited tissue repertoire of poliovirus infection and the manifestation of neurovirulence as discussed below.

5 Shut-Off of Host Protein Synthesis After Poliovirus Infection

The precipitous and drastic reduction of host protein synthesis after poliovirus infection is one of the first described virus-induced shut-off phenomena (for reviews see EHRENFELD 1984; SONENBERG 1987; CARRASCO and CASTRILLO 1987). The ability of poliovirus to cause a strong shut-off of host protein synthesis provides poliovirus mRNA facilitated access to the host translational machinery by avoiding competition from host mRNAs.

What is the mechanism of shut-off of host protein synthesis by poliovirus? Most evidence points to the inactivation of eIF-4F as the cause of the shut-off. When purified from infected cells eIF-4F is inactive in an in vitro reconstituted translation system (ETCHISON et al. 1984). All other initiation factors could be purified from poliovirus-infected cells and shown to be active in reconstituted translation systems (DUNCAN et al. 1983; ETCHISON et al. 1984). These findings are consistent with the results showing that addition of eIF-4F could restore the translation of capped mRNAs in extracts prepared from poliovirus-infected cells (TAHARA et al. 1981; EDERY et al. 1984). The inactivation of eIF-4F is also consistent with the findings that eIF-4A and eIF-4B in HeLa extracts were incapable of cross-linking to the mRNA 5' cap structure, in spite of their functionality (LEE and SONENBERG 1982; PELLETIER and SONEBERG 1985a). This could be explained if eIF-4F is inactivated, since cross-linking of eIF-4A and eIF-4B to the mRNA cap structure is dependent on the prior interaction of eIF-4F with the cap structure (EDERY et al. 1983). As predicted, if eIF-4F were inactivated in the infected extract, addition of exogenous eIF-4F rescued the ability of eIF-4B to cross-link to the mRNA cap structure (LEE et al. 1985a).

How is eIF-4F inactivated? The only clear and specific modification of eIF-4F occurs to p220. This subunit is proteolyzed in poliovirus-infected cells to yield three to four polypeptides of 110–130 kDa, as determined first by immuno-blotting (ETCHISON et al. 1982; LEE et al. 1985b). Furthermore, a modified eIF-4F containing the proteolytic cleavage products and eIF-4E was purified from poliovirus-infected cells (LEE et al. 1985a; BUCKLEY and EHRENFELD 1987). It does not appear that the 24 kDa subunit (eIF-4E) of eIF-4F is modified in poliovirus-infected cells (LEE and SONENBERG 1982; BUCKLEY and EHRENFELD 1986). Taken together, these findings suggest that in poliovirus-infected cells the initial defect in translation of cellular capped mRNAs is the interaction of eIF-4F with the mRNA cap structure. This is consistent with the finding that cross-linking of eIF-4E to the cap structure is significantly reduced in preparations from poliovirus-infected cells (LEE and SONENBERG 1982). It was also found that eIF-4E from poliovirus-infected cells sedimented as a low-molecular weight species in contrast to its sedimentation as a high molecular weight complex in mock-infected cells (HANSEN et al. 1982).

What is the effector of the p220 cleavage? Several experiments point to the poliovirus-coded protease 2A (2A^{pro}) as the mediator of p220 cleavage. 2A^{pro} is responsible for the cleavage of two Tyr-Gly amino acid pairs: one between VP1 and 2A and the second between 3C′ and 3D′ (Toyoda et al. 1986). A mutated virus containing an insertion of the amino acid leucine in 2A^{pro} gave rise to a small-plaque phenotype in HeLa and CV-1 cells (Bernstein et al. 1985). Moreover, in CV-1 cells the early inhibition of host protein synthesis did not occur. In HeLa cells a general inhibition of protein synthesis, including that of poliovirus was observed, and might have been due to the increased phosphorylation of the α-subunit of eIF-2 (Bernstein 1988). Most striking is the finding that in both HeLa and CV-1 cells, p220 was not cleaved following infection with the mutant 2A^{pro} virus (Bernstein et al. 1985). The lack of p220 cleavage cannot be attributed to the reduced levels of poliovirus proteins in the infected cells, since cleavage of p220 was even observed in cells infected with poliovirus in the presence of guanidine, in which synthesis of poliovirus protein is undetected (Bonneau and Sonenberg 1987). In support of the involvement of 2A^{pro} in the cleavage of p220, Racaniello and colleagues (Dolph et al. 1987; O'Neill and Racaniello 1989) generated a different insertional mutation in 2A^{pro}, which also resulted in a small plaque mutant. However, 2A^{pro} most probably does not cleave p220 directly. Antibodies directed against 2A^{pro} did not inhibit in vitro proteolysis of p220 when the latter was incubated with extracts from poliovirus-infected cells, and p220 proteolyzing activity from poliovirus-infected cells did not copurify with 2A^{pro} (Lloyd et al. 1986). In a more recent study it was shown that translation of 2A^{pro} mRNA in a HeLa extract caused cleavage of p220, but antibody against 2A^{pro} inhibited cleavage when present during translation, but not when added after translation of 2A^{pro} mRNA (Kräusslich et al. 1987; Lloyd et al. 1988). Thus, these results provide direct evidence for the role of 2A^{pro} as a mediator of p220 cleavage resulting in inactivation of eIF-4F and cessation of cellular protein synthesis. The putative cellular protease that cleaves p220 has not been identified. However, recent experiments demonstrated that p220 is a specific substrate for calpain, which is a cytoplasmic thiol, Ca^{2+}-activated protease (E. Ehrenfeld, personal communication). It would be of interest to determine whether 2A^{pro} activates calpain.

Cleavage of p220 is necessary, but not sufficient to cause complete inhibition of host protein synthesis. When poliovirus infection was performed in the presence of guanidine or 3-methyl quercetin, which strongly inhibit poliovirus RNA synthesis resulting in minimal synthesis of poliovirus protein, p220 is completely cleaved (Bonneau and Sonenberg 1987; Lloyd et al. 1987). However, translation of cellular mRNAs is reduced only to approximately 30% of control levels (Bonneau and Sonenberg 1987). This suggests firstly that translation of celluar mRNA in vivo can occur at a reduced level in a cap-independent fashion. Therefore, there must be a second event occurring during poliovirus infection that inhibits cellular cap-independent translation that is important for the complete abrogation of cellular translation. This event is unlikely to be competition between poliovirus mRNA and cellular mRNAs for a limiting translation factor. This is concluded from results showing that translation of globin mRNA in micrococcal nuclease-treated extracts prepared from poliovirus-infected cells (and therefore evidently not containing intact poliovirus RNA) was completely abolished (BONNEAU and SONENBERG 1987). Since poliovirus RNA was degraded by the nuclease treatment, competition could not explain the reduced globin translation.

5.1 eIF-2α Phosphorylation in Poliovirus-Infected Cells

What might be this second event? A likely candidate is the phosphorylation of the α-subunit of eIF-2. A study by BLACK et al. (1989) demonstrated that during poliovirus infection, double-stranded RNA activated kinase (dsI) becomes autophosphorylated and activated. There is also significant degradation of dsI during the infection. The only known substrate of dsI in vivo besides itself is eIF-2α (ROBERTS et al. 1976) and its phosphorylation results in inhibition of protein synthesis. In accordance with dsI activation, phosphorylation of eIF-2α increased approximately threefold relative to mock-infected cells between 3 and 5 h postinfection (BLACK et al. 1989). Control experiments ruled out the possibility that phosphorylation of eIF-2α occurred in vitro after cell disruption (BLACK et al. 1989). Similar results were obtained by O'NEILL and RACANIELLO (1989). In contrast to these results, RANSONE and DASGUPTA (1987, 1988) reported that despite the activation of the protein kinase, they have failed to detect an increase in eIF-2α phosphorylation in poliovirus-infected cells. In addition, RANSONE and DASGUPTA (1988) identified a heat-sensitive inhibitor that selectively blocked the dsI-induced phosphorylation of the α-subunit of eIF-2. There are several possible explanations for the discrepancy in the results, as pointed by BLACK et al. (1989). These include the possibilities of phosphatase activation that dephosphorylates eIF-2α during cell extraction and changes in subcellular location of eIF-2 during infection. However, it is also possible that under certain conditions an eIF-2α phosphorylation inhibitor is activated in poliovirus-infected cells to control the extent of phosphorylation (RANSONE and DASGUPTA 1988). In fact, the operative mechanism to control the level of eIF-2α phosphorylation is by degrading the activated dsI in poliovirus-infected cells (BLACK et al. 1989). A similar phenomenon was reported in cells infected with EMCV, and vaccinia virus (HOVANESSIAN et al. 1987). Thus, several mechanisms might operate in poliovirus infected cells to regulate eIF-2α activity. It would be of interest to determine if 2A^pro, which mediates the proteolysis of p220, is also implicated in the degradation of dsI.

Does eIF-2α phosphorylation have a differential effect on poliovirus versus cellular mRNA translation? Maximum phosphorylation of eIF-2α occurs at 3 h after poliovirus infection, when poliovirus translation is maximal (BLACK et al. 1989) and when no translation of cellular mRNAs is usually detectable (BONNEAU and SONENBERG 1987). An attractive possibility to explain the differential ability of poliovirus mRNA to translate under these conditions is based on the hypothesis that activation of dsI is localized, namely that only the translation of the dsI, inducing mRNA will be inhibited. Several reports support the idea that activation of dsI is localized (DeBenedetti and Baglioni 1984; Kaufman and Murtha 1987). In poliovirus-infected cells the secondary structure in the 5′ noncoding regions of the cellular mRNAs cannot be unwound because of the inactivation of eIF-4F. Consequently, these double-stranded regions of the mRNA could activate dsI, and cause the *trans*-inhibition of cellular mRNA translation including cap-independent translation. A precedent for *trans*-inhibition of mRNA translation by the secondary structure of mRNA was reported for HIV-1 (Edery et al. 1989; Sengupta and Silverman 1989). In this case a stable stem and loop structure that is present at the 5′ end of HIV-1 mRNAs inhibited in vitro translation in *trans* by activating dsI which in turn phosphorylated the α-subunit of eIF-2. In other studies (Pratt et al. 1988) it was shown that preparations of poly A$^+$ mRNA from cells also had a *trans*-inhibitory effect.

6 Poliovirus Translation and Neurovirulence

Specific nucleotides in the 5′ UTR of poliovirus RNAs have been correlated with neurovirulence. All three poliovirus serotypes have major neurovirulence attenuating mutations in the 5′ UTR. In Sabin type 3 a base change at position 472 (C to U) in addition to a C to U mutation at nucleotide 2034 determine the attenuation phenotype (Evans et al. 1985; Westrop et al. 1989). In the case of the Sabin type 1 strain, there are multiple attenuating mutations, but an important mutation was mapped to position 480 (Omata et al. 1986; Nomoto et al. 1987; Kawamura et al. 1989). It has been suggested that a similar change (from A to G) at position 487 of Sabin 2 is also responsible for the acquisition of neurovirulence (Minor and Dunn 1988). Consistent with this suggestion a fragment of P2 containing nucleotides 456 to 628 conferred an attenuated phenotype (Moss et al. 1989). Since the 5′ UTR of poliovirus RNA is involved in translation initiation, it is likely that the attenuation phenotype is the outcome of a modified translational behavior. This notion is strongly supported by the studies of Agol and collaborators (Svitkin et al. 1985, 1988), who showed first that translation of mRNAs from attenuated strains of poliovirus types 1 and 3 is less efficient than that of their neurovirulent counterparts. In addition, Svitkin et al. (1988) demonstrated that the initiation-correcting factor (ICF; see above) was less efficient in correcting the aberrant translation pattern in reticulocyte lysate of attenuated viral RNAs than of neurovirulent RNAs. Further evidence for the correlation between attenuation and translational efficiency was obtained from experiments in tissue culture by using human neuroblastoma cell lines. Recombinant poliovirus that contains the 5′ UTR of Sabin 3 with the mutation of C to U at position 472 replicated to a tenfold-lower titer than the neurovirulent virus, and the defect was determined to be at the translational level (LaMonica and Racaniello 1989). Similarly, Agol et al. (1989) found that determinants in the 5′ UTR of Sabin strains 1 and 2 are responsible for the inability of these strains to replicate in human neuroblastoma cell lines. Attenuated poliovirus can replicate efficiently in HeLa cells (LaMonica and Racaniello 1989) and therefore it is an attractive idea that a translational factor is more limiting in neural cells than in HeLa cells. This is consistent with the idea that the 5′ UTR of the poliovirus genome contains tissue-specific *cis*-acting elements, and in conjunction with tissue-specific *trans*-acting factors (such as p52) are responsible for the tissue specificity of poliovirus. It would be of interest to determine if p52 is indeed implicated in the differential expression of attenuated and neurovirulent poliovirus in neural cells.

7 Conclusions and Future Work

The work conducted in the past few years on picornavirus RNA translation has engendered fundamental knowledge about this critical step in virus replication. A key finding was that ribosome entry on picornavirus RNAs does not occur from the 5′ end of RNA, as was postulated for all eukaryotic mRNAs, but rather by internal binding to a sequence in the 5′ noncoding region. This mechanism of ribosome binding answers long-standing questions about the ability of poliovirus RNA to translate without the cap structure, and in poliovirus-infected cells at a time when cellular mRNAs cannot be translated.

Future work will concentrate on the structural requirements for the RLP and IRES elements and the *trans*-acting factor(s) that promote this process. Efforts will continue to elucidate the details of the mechanism of the shut-off of host protein synthesis after poliovirus infection. In particular, it is critically important to identify the protease that cleaves the p220 component of eIF-4F. It became clear however, that in addition to p220, phosphorylation of eIF-2α could also play an important role in the shut-off of host protein synthesis. Studies are in progress to clarify the mechanism by which poliovirus counteracts the biochemical pathway activated by double-stranded RNA that ultimately results in eIF-2α phosphorylation. An intriguing possibility is that 2A^{pro} is also responsible for the degradation of dsI.

Of considerable importance is the relationship between tissue-specific translation of poliovirus RNA and neurovirulence. The recent experiments from Racaniello's and Agol's laboratories clearly demonstrate a correlation between neurovirulence and translation in a neuroblastoma cell line, but not in HeLa cells. An attractive hypothesis is that differences in translation factors between tissues account for such differential translation. It is anticipated that the invigorated research effort on picornavirus translation will continue with a vengeance. These studies might yield new surprises and insights into the complex machinery of protein synthesis in eukaryotic cells, and should shed light on the control of translation during normal cell growth and in disease.

Acknowledgements. I am grateful to my colleagues and collaborators who carried out this work and to the members of my laboratory for their comments on the manuscript.

References

Abramson RD, Dever TE, Lawson TG, Ray BK, Thach RE, Merrick WC (1987) The ATP-dependent interaction of eukaryotic initiation factors with mRNA. J Biol Chem 262: 3826–3832

Agol VI, Drozdov SG, Ivannikova TA, Kolesnikova MS, Korolev MB, Tolskaya EA (1989) Restricted growth of attenuated poliovirus strains in cultured cells of a human neuroblastoma. J Virol 63: 4034–4038

Alonso MA, Carrasco L (1981) Reversion by hypotonic medium of the shutoff of protein synthesis induced by encephalomyocarditis virus. J Virol 37: 535–540

Bablanian R, Russell WC (1974) Adenopolypeptide synthesis in the presence of non-replicating poliovirus. J Gen Virol 24: 261–279

Banerjee AK (1980) 5'-Terminal cap structure in eukaryotic messenger ribonucleic acids. Microbiol Rev 44: 175–205

Bernstein HD, Sonenberg N, Baltimore D (1985) Poliovirus mutant that does not selectively inhibit host cell protein synthesis. Mol Cell Biol 50: 2913–2923

Bernstein HD (1987) Characterization of poliovirus mutants derived from an infectious cDNA clone. Ph.D. Thesis, Massachussets Institute of Technology

Bienkowska-Szewczyk K, Ehrenfeld E (1988) An internal 5'-noncoding region required for translation of poliovirus RNA in vitro. J Virol 62: 3068–3072

Black TL, Safer B, Hovanessian A, Katze M (1989) The cellular 68,000 Mr protein kinase is highly phosphorylated and activated yet significantly degraded during poliovirus infection: implications for translation regulation. J Virol 63: 2244–2251

Bonneau A-M, Sonenberg N (1987) Proteolysis of the p220 component of the cap-binding protein complex is not sufficient for complete inhibition of host cell protein synthesis after poliovirus infection. J Virol 61: 986–991

Brown BA, Ehrenfeld E (1979) Translation of poliovirus RNA in vitro: changes in cleavage pattern and initiation sites by ribosomal salt wash. Virology 97: 396–405

Buckley B, Ehrenfeld E (1986) Two-dimensional gel analyses of the 24-kDa cap binding protein from poliovirus-infected and uninfected HeLa cells. Virology 152: 497–501

Buckley B, Ehrenfeld E (1987) The cap-binding protein complex in uninfected and poliovirus-infected HeLa cells. J Biol Chem 262: 13599–13606

Carrasco L, Castrillo JL (1987) The regulation of translation in picornavirus-infected cells. In: Carrasco L (ed) Mechanism of viral toxicity in animal cells. CRC, Boca Raton, pp 116–146

Castrillo JL, Carrasco L (1988) Adenovirus late protein synthesis is resistant to the inhibition of translation induced by poliovirus. J Biol Chem 262: 7328–7334

Chang L-J, Pryciak P, Ganem D, Varmus HE (1989) Biosynthesis of the reverse transcriptase of hepatitis B viruses involves de novo translational initiation not ribosomal frameshifting. Nature 337: 364–368

Curran J, Kolakofsky D (1989) Scanning independent ribosomal initiation of the Sendai virus Y proteins in vitro and in vivo. EMBO J 8: 521–526

Daniels-McQueen S, Detjen BM, Grifo JA, Merrick WC, Thach RE (1983) Unusual requirements for optimal translation of polioviral RNA in vitro. J Biol Chem 258: 7195–7199

deBenedetti A, Baglioni C (1984) Inhibition of mRNA binding to ribosomes by localized activation of dsRNA-dependent protein kinase. Nature 311: 79–81

del Angel RM, Papavassiliou AG, Fernandez-Tomas C, Silverstein SJ, Racaniello VR (1989) Cell proteins bind to multiple sites within the 5'-untranslated region of poliovirus RNA. Proc Natl Acad Sci USA 86: 8299–8303

Detjen BM, Jen G, Thach RE (1981) Encephalomyocarditis viral RNA can be translated under conditions of polio-induced translation shut-off in vivo. J Virol 38: 777–781

Dolph PJ, Racaniello VR, Villamarin A, Palladino F, Schneider RJ (1988) The adenovirus tripartite leader may eliminate the requirement for cap-binding protein complex during translation initiation. J Virol 62: 2059–2066

Dorner AJ, Semler BJ, Jackson RJ, Hanecak R, Duprey RE, Wimmer E (1984) In vitro translation of poliovirus RNA: utilization of internal initiation sites in reticulocyte lysate. J Virol 50: 507–514

Duncan R, Etchison E, Hershey JWB (1983) Protein synthesis initiation factors eIF-4A and eIF-4B are not altered by poliovirus infection of HeLa cells. J Biol Chem 258: 7236–7239

Edery I, Sonenberg N (1985) Cap-dependent RNA splicing in a HeLa nuclear extract. Proc Natl Acad Sci USA 82: 7590–7594

Edery I, Humbelin M, Darveau D, Lee KAW, Milburn S, Hershey JWB, Trachsel H, Sonenberg N (1983) Involvement of eIF-4A in the cap recognition process. J. Biol Chem 258: 11398–11403

Edery I, Lee KAW, Sonenberg N (1984) Functional characterization of eukaryotic mRNA cap binding protein complex: effects on translation of capped and naturally uncapped RNAs. Biochemistry 23: 2456–2462

Edery I, Pelletier J, Sonerberg N (1987) Regulation of protein synthesis by cap binding protein complex. In: Ilan J (ed) Translational regulation of gene expression. Plenum p 335 New York

Edery I, Petryshyn R, Sonenberg N (1989) Activation of double-stranded RNA-dependent kinase (dsI) by the TAR region of HIV-1 mRNA: a novel translational control mechanism. Cell 56: 303–312

Ehrenfeld E (1984) Picornavirus inhibition of host cell protein synthesis. In: Fraenkel-Conrat H, Wagner RR (eds) Comprehensive virology, vol 19. Plenum, New York, pp 177–221

Etchison D, Milburn SC, Edery I, Sonenberg N, Hershey JWB (1982) Inhibition of HeLa cell protein synthesis following poliovirus infection correlates with the proteolysis of a 220,000 dalton polypeptide associated with eukaryotic initiation factor 3 and a cap binding protein complex. J Biol Chem 257: 14806–14810

Etchison D, Hansen J, Ehrenfeld E, Edery I, Sonenberg N, Milburn SC, Hershey JWB (1984) Demonstration in vitro that eucaryotic initiation factor 3 is active but that a cap-binding protein complex is inactive in poliovirus infected HeLa cells. J Virol 51: 832–837

Evans DMA, Dunn G, Minor PD, Schild GC, Cann AJ, Stanway G, Almond JW, Currey K, Maizel JV Jr (1985) Increased neurovirulence associated with a single nucleotide change in a noncoding region of the Sabin type 3 poliovaccine genome. Nature 314: 548–553

Fernandez-Munoz R, Darnell JE (1976) Structural difference between the 5' termini of viral and cellular mRNA in poliovirus-infected cells: possible basis for the inhibition of host protein synthesis. J Virol 18: 719–726

Flanagan JB, Petterson RF, Ambros V, Hewlett MJ, Baltimore D (1977) Covalent linkage of a protein to a defined nucleotide sequence at the 5'-terminus of virion and replicative intermediate RNAs of poliovirus. Proc Natl Acad Sci USA 74: 961–965

Furuichi Y, LaFiandra A, Shatkin AJ (1977) 5'-Terminal structure and mRNA stability. Nature 266: 235–239

Gehrke L, Auron PE, Quigley GJ, Rich A, Sonenberg N (1983) 5'-Conformation of capped alfalfa mosaic virus ribonucleic acid 4 may reflect its independence of the cap structure or of cap binding protein for efficient translation. Biochemistry 22: 5157–5164

Georgiev O, Mous J, Birnstiel ML (1984) Processing and nucleo-cytoplasmic transport of histone gene transcripts. Nucleic Acids Res 12: 8539–8551

Green MR, Maniatis T, Melton DA (1983) Human β-globin pre-mRNA synthesized in vitro is accurately spliced in Xenopus oocyte. Cell 31: 681–694

Grifo JA, Tahara SM, Leis JP, Morgan MA, Shatkin AJ, Merrick WC (1982) Characterization of eukaryotic initiation factor 4A, a protein involved in ATP-binding of globin mRNA. J Biol Chem 257: 5246–5252

Grifo JA, Tahara SM, Morgan MA, Shatkin AJ, Merrick WC (1983) New initiation factor activity required for globin mRNA translation. J Biol Chem 258: 5804–5810

Hansen J, Etchison D, Hershey JWB, Ehrenfeld E (1982) Association of cap-binding protein with eucaryotic initiation factor 3 in initiation factor preparations from uninfected and poliovirus-infected HeLa cells. J Virol 42: 200–207

Hassin D, Korn R, Horwitz MS (1986) A major internal initiation site for the in vitro translation of the adenovirus DNA polymerase. Virology 155: 214–224

Herman RC (1986) Internal initiation of translation on the vesicular stomatitis virus phosphoprotein mRNA yields a second protein. J Virol 58: 797–804

Hewlett MJ, Rose JK, Baltimore D (1976) 5'-Terminal structure of poliovirus polyribosomal RNA is pUp. Proc Natl Acad Sci USA 73: 327–330

Hiremath LA, Hiremath ST, Rychlik W, Joshi S, Domier LL, Rhoads RE (1989) In vitro synthesis, phosphorylation, and localization on 48S initiation complexes of human protein synthesis initiation factor 4E. J Biol Chem 264: 1132–1138

Holland JJ (1961) Receptor affinities as major determinants of enterovirus tissue tropism in humans. Virology 15: 312–326

Hovanessian AG, Galabru J, Meurs E, Buffet-Janvresse C, Svab J, Robert N (1987) Rapid decrease in the levels of the double stranded RNA-dependent protein kinase during virus infections. Virology 159: 126–136

Jagus R, Anderson WF, Safer B (1981) The regulation of initiation of mammalian protein synthesis. Prog Nucleic Acid Res Mol Biol 25: 127–185

Jang SK, Kräusslich H-G, Nicklin MJH, Duke GM, Palmenberg AC, Wimmer E (1988) A segment of the 5' non-translated region of encephalomyocarditis virus RNA directs internal entry of ribosomes during in vitro translation. J Virol 62: 2636–2643

Jang SK, Davies MV, Kaufman RJ, Wimmer E (1989) Initiation of protein synthesis by internal entry of ribosomes into the 5' non-translated region of encephalomyocarditis virus RNA in vivo. J Virol 63: 1651–1660

Kaufman RJ, Murtha P (1987) Translational control mediated by eucaryotic initiation factor-2 is restricted to specific mRNAs in transfected cells. Mol Cell Biol 7: 1568–1571

Kawamura N, Kohara M, Abe S, Komatsu T, Tago K, Arita M, Nomoto A (1989) Determinants in the 5' noncoding region of poliovirus Sabin 1 RNA that influence the attenuation phenotype. J Virol 63: 1302–1309

Kitamura N, Semler BL, Rothberg PG, Larsen GR, Adler CJ, Dorner AJ, Emini EA, Henecak R, Lee JJ, van der Werf S, Anderson CW, Wimmer E (1981) Primary structure, gene organization and polypeptide expression of poliovirus RNA. Nature 291: 547–553

Konarska MM, Padgett RA, Sharp PA (1984) Recognition of cap structure in splicing in vitro of mRNA precursors. Cell 38: 731–736

Kozak M (1983) Comparison of initiation of protein synthesis in procaryotes, eucaryotes, and organelles. Microbiol Rev 47: 1–45

Kräusslich H-G, Nicklin MGH, Toyoda H, Etchison D, Wimmer E (1987) Poliovirus proteinase 2A induces cleavage of eukaryotic initiation factor 4F polypeptide p220. J Virol 61: 2711–2718

Kuge S, Nomoto N (1987) Construction of viable deletion and insertion mutants of the Sabin strain of type 1 poliovirus: function of the 5' noncoding sequence in viral replication. J Virol 61: 1478–1487

Kuge S, Kawamura N, Nomoto A (1989) Genetic variation occurring on the genome of an in vitro insertion mutant of poliovirus type 1. J Virol 63: 1069–1075

La Monica N, Racaniello VR (1989) Differences in replication of attenuated and neurovirulent polioviruses in human neuroblastoma cell line SH-SYSY. J Virol 63: 2357–2360

La Monica N, Meriam C, Racaniello VR (1986) Mapping of sequences required for mouse neurovirulence of poliovirus type 2 Lansing. J Virol 57: 515–525

Lawson TG, Lee KA, Maimone MM, Abramson RD, Dever TE, Merrick WC, Thach RE (1989) ATP-dependent dissociation of helical structures in mRNA molecules by eukaryotic initiation factors -4F, -4A and -4B. Biochemistry 28: 4729–4734

Lee KAW, Sonenberg N (1982) Inactivation of cap-binding proteins accompanies the shut-off of host protein synthesis by poliovirus. Proc Natl Acad Sci USA 79: 3447–3451

Lee KAW, Edery I, Sonenberg N (1985a) Isolation and structural characterization of cap-binding proteins from poliovirus-infected HeLa cells. J Virol 54: 515–524

Lee KAW, Edery I, Hanecak R, Wimmer E, Sonenberg N (1985b) Poliovirus protease 3C (P3-7c) does not cleave P220 of the eukaryotic mRNA cap-binding protein complex. J Virol 55: 489–493

Lee YF, Nomoto A, Detjen BM, Wimmer E (1977) A protein covalently linked to poliovirus genome RNA. Proc Natl Acad Sci USA 74: 59–63

Linder P, Lasko PF, Ashburner M, Leroy P, Nielsen PJ, Nishi K, Schnier J, Slonimski P (1989) Birth of the D-E-A-D box. Nature 337: 121–122

Lloyd RE, Toyoda H, Etchison D, Wimmer E, Ehrenfeld E (1986) Cleavage of the cap-binding protein complex polypeptide p220 is not effected by the second poliovirus protease 2A. Virology 150: 299–303

Lloyd RE, Jense HG, Ehrenfeld E (1987) Restriction of translation of capped mRNA in vitro as a model for poliovirus induced inhibition of host cell protein synthesis: relationship to p220 cleavage. J Virol 61: 2480–2488

Lloyd RE, Grubman MJ, Ehrenfeld E (1988) Relationship of p220 cleavage during picornavirus infection to 2A proteinase sequences. J Virol 62: 4216–4223

McClain K, Stewart M, Sullivan M, Maizel JV Jr (1981) Ribosomal binding sites on poliovirus RNA. Virology 113: 150–167

McCormick W, Penman S (1968) Replication of mengovirus in HeLa cells preinfected with non-replicating poliovirus. J Virol 2: 859–864

Meerovitch K, Pelletier J, Sonenberg N (1989) A cellular protein that binds to the 5'-noncoding region of poliovirus RNA: implications for internal translation initiation. Genes Dev 3: 1026–1034

Merrick WC, Abramson RD, Anthony DD Jr. Dever TE, Caliendo AM (1987) Involvement of nucleotides in protein synthesis initiation. In: Jlan J (ed) Translational regulation of gene expression. Plenum, New York, p 265

Minor PD, Dunn G (1988) The effect of sequences in the 5'-noncoding region on the replication of polioviruses in the human gut. J Gen Virol 69: 1091–1096

Moldave K (1985) Eukaryotic protein synthesis. Annu Rev Biochem 54: 1109–1149

Moss EG, O'Neill RE, Racaniello VR (1989) Mapping of attenuating sequences of an avirulent poliovirus type 2 strain. J Virol 63: 1884–1890

Munoz A, Alonso MA, Carrasco L (1984) Synthesis of heat-shock proteins in HeLa cells: inhibition by virus infection. Virology 137: 150–159

Nicklin MJH, Kräusslich HG, Toyoda H, Dunn JJ, Wimmer E (1987) Poliovirus polypeptide precursors: expression in vitro and processing by exogenous 3C and 2A proteinases. Proc Natl Acad Sci USA 84: 4002–4006

Nomoto A, Lee YF, Wimmer E (1976) The 5' end of poliovirus mRNA is not capped with $m^7G(5')ppp(5')Np$. Proc Natl Acad Sci USA 73: 375–380

Nomoto A, Kohara M, Kuge S, Kawamura N, Arita M, Komatsu T, Abe S, Semler, BL, Wimmer E, Itoh H (1987) In: Brinton MA, Rueckert R (eds) Positive strand RNA viruses. Liss, New York, p 437 (UCLA symposia on molecular and cellular biology, vol 24)

Okada Y, Toda G, Oka H, Nomoto A, Yoshikura H (1987) Poliovirus infection of established human blood cell lines: relationship between differentiation stage and susceptibility or cell killing. Virology 156: 238–245

Omata T, Kohara M, Kuge S, Komatsu T, Abe S, Semler BL, Kameda B, Itoh H, Arita H, Wimmer E, Nomoto A (1986) Genetic analysis of the attenuation phenotype of poliovirus type 1. J Virol 58: 348–358

O'Neill R, Racaniello VR (1989) Inhibition of translation in cells infected with a poliovirus 2A^{pro} mutant correlates with phosphorylation of the alpha subunit of eucaryotic initiation factor 2. J Virol 63: 5069–5075

Pain VM (1986) Initiation of protein synthesis in mammalian cells. Biochem J 235: 625–637

Pelletier J, Sonenberg N (1985a) Photochemical cross-linking of cap binding proteins to eucaryotic mRNAs. Effect of mRNA 5' secondary structure. Mol Cell Biol 5: 3222–3230

Pelletier J, Sonenberg N (1985b) Insertion mutagenesis to increase secondary structure within the 5' noncoding region of a eukaryotic mRNA reduces translational efficiency. Cell 40: 515–526

Pelletier J, Sonenberg N (1988) Internal initiation of translation of eukaryotic mRNA directed by a sequence derived from poliovirus RNA. Nature 334: 320–325

Pelletier J, Kaplan G, Racaniello VR, Sonenberg N (1988a) Cap-independent translation of poliovirus mRNA is conferred by sequence elements wthin the 5' noncoding region. Mol Cell Biol 8: 1103–1112

Pelletier J, Kaplan G, Racaniello VR, Sonenberg N (1988b) Translational efficiency of poliovirus mRNA: mapping of inhibitory cis-acting elements within the 5' noncoding region. J Virol 62: 2219–2227

Pelletier J, Flynn ME, Kaplan G, Racaniello VR, Sonenberg N (1988c) Mutational analysis of upstream AUG codons of poliovirus RNA. J Virol 62: 4486–4492

Pelletier J, Sonenberg N (1989) Internal binding of eucaryotic ribosomes on poliovirus RNA: translation in HeLa cell extracts. J Virol 63: 441–444

Perez-Bercoff R (1982) But is the 5' end of messenger RNA always involved in initiation? In: Perez-Bercoff R (ed) Protein synthesis in eukaryotes. Plenum, New York, pp 242–252

Phillips BA, Emmert A (1986) Modulation of the expression of poliovirus proteins in reticulocyte lysates. Virology 148: 255–267

Pilipenko EM, Blinov VM, Romanova LI, Sinyakov AN, Maslova SV, Agol VI (1989a) Conserved structural domains in the 5'-untranslated region of picornaviral genomes: an anlysis of the segment controlling translation and neurovirulence. Virology 168: 201–209

Pilipenko EM, Blinov VM, Chernov BK, Dmitrieva TM, Agol VI (1989b) Conservation of the secondary structure element of the 5'-untranslated region of cardio- and aphtovirus RNAs. Nucl Acids Res 17: 5701–5711

Pratt G, Galpine A, Sharp N, Palmer S, Clemens MJ (1988) Regulation of in vitro translation by double-stranded RNA in mammalian cell RNA preparations. Nucleic Acids Res 16: 3497–3510

Proud CG (1986) Guanine nucleotides, protein phosphorylation and the control of translation. Trends Biochem Sci 11: 73–77

Racaniello VR, Baltimore D (1981) Molecular cloning of poliovirus cDNA and determination of the complete nucleotide sequence of the viral genome. Proc Natl Acad Sci USA 78: 4887–4891

Ransone LJ, Dasgupta A (1987) Activation of double-stranded RNA-activated protein kinase in HeLa cells after poliovirus infection does not results in increased phosphorylation of eucaryotic initiation factor-2. J Virol 61: 1781–1787

Ransone LJ, Dasgupta A (1988) A heat sensitive inhibitor in poliovirus-infected cells which selectively blocks phosphorylation of the α subunit of eucaryotic initiation factor 2 by the double-stranded RNA-activated protein kinase. J Virol 62: 3551–3557

Ray BK, Lawson TG, Kramer JC, Cladaras MH, Grifo JA, Abramson RD, Merrick WC, Thach RE (1985) ATP-dependent unwinding of messenger RNA structure by eukaryotic initiation factors. J Biol Chem 260: 7651–7658

Rivera VM, Welsh JD, Maizel V Jr (1988) Comparative sequence analysis of the 5' noncoding region of enteroviruses and rhinoviruses. Virology 165: 42–50

Roberts WK, Hovanessian A, Brown RE, Clemens MJ, Kerr IM (1976) Interferon-mediated protein kinase and low-molecular weight inhibitor of protein synthesis. Nature 264: 477–480

Rose JK, Trachsel H, Leong K, Baltimore D (1978) Inhibition of translation by poliovirus: inactivation of a specific initiation factor. Proc Natl Acad Sci USA 75: 2732–2736

Rozen F, Pelletier J, Trachsel H, Sonenberg N (1989) A lysine substitution in the ATP-binding site of eucaryotic initiation factor 4A abrogates nucleotide-binding activity. Mol Cell Biol 9: 4061–4063

Rozen F, Edery I, Meerovitch K, Dever TE, Merrick WC, Sonenberg N (1990) Bidirectional RNA helicase activity of eucaryotic translation initiation factors -4A and -4F. Mol Cell Biol 10: 1134–1144

Safer B (1983) 2B or not 2B: regulation of catalytic utilization of eIF-2. Cell 33: 7–8

Sarkar G, Edery I, Sonenberg N (1985) Photoaffinity labelling of the cap-binding protein complex with ATP/dATP. J Biol Chem 260: 13831–13837

Sarnow P (1989) Translation of glucose-regulated protein 78/immunoglobulin heavy-chain binding protein mRNA is increased in poliovirus-infected cells at a time when cap-dependent translation of cellular mRNAs is inhibited. Proc Natl Acad Sci USA 86: 5795–5799

Schlicht H-J, Radziwill G, Schaller H (1989) Synthesis and encapsidation of duck hepatis B virus reverse transcriptase do not require formation of core-polymerase fusion proteins. Cell 56: 85–92

SenGupta DN, Silverman RH (1989) Activation of interferon regulated, dsRNA-dependent enzymes by human immunodeficiency virus-1 leader RNA. Nucleic Acids Res 17: 969–978

Shatkin AJ (1976) Caping of eucaryotic mRNAs. Cell 9: 645–653

Shatkin AJ (1985) mRNA Cap binding proteins: essential factors for initiating translation. Cell 40: 223–224

Shih DS, Shih CT, Kew O, Pallansch M, Rueckert RR, Kaesberg P (1978) Cell-free synthesis and processing of the proteins of poliovirus. Proc Natl Acad Sci USA 75: 5807–5811

Shih DS, Park I-W, Evans CL, Jaynes JM, Palmenberg AC (1987) Effects of cDNA hybridization on translation of encephalomyocarditis virus RNA. J Virol 61: 2033–2037

Shirman GA, Maslova SV, Gavrilovskaya IN, Agol VI (1973) Stimulation of restricted reproduction of EMC virus in HeLa cells by non-replicating poliovirus. Virology 51: 1–10

Skinner MA, Racaniello VR, Dunn G, Cooper J, Minor PD, Almond JW (1989) New model for the secondary structure of the 5' non-coding RNA of poliovirus is supported by biochemical and genetic data that also show that RNA secondary structure is important in neurovirulence. J Mol Biol 207: 379–392

Sonenberg N (1981) ATP/Mg^{++}-dependent cross-linking of cap binding proteins to the 5' end of eukaryotic mRNA. Nucleic Acids Res 9: 1643–1656

Sonenberg N (1987) Regulation of translation by poliovirus. Adv Virus Res 33: 175–204

Sonenberg N (1988) Cap-binding proteins of eukaryotic messenger RNA: functions in initiation and control of translation. Prog Nucleic Acid Res Mol Biol 35: 173–207

Sonenberg N, Merrick WC, Morgan MA, Shatkin AJ, (1978) A polypeptide in eukaryotic initiation factors that cross-links specifically to the 5' terminal cap mRNA. Proc Natl Acad Sci USA 75: 4843–4847

Sonenberg N, Guertin D, Cleveland D, Trachsel H (1981) Probing the function of the eukaryotic 5′ cap structure by using a monoclonal antibody directed against cap-binding proteins. Cell 27: 563–572

Sonenberg N, Guertin D, Lee KAW (1982) Capped mRNAs with reduced secondary structure can function in extracts from poliovirus-infected cells. Mol Cell Biol 2: 1633–1638

Svitkin YV, Maslova SV, Agol VI (1985) The genomes of attenuated and virulent poliovirus strains differ in their in vitro translation efficiencies. Virology 147: 243–252

Svitkin YV, Pestova TV, Maslova SV, Agol VI (1988) Point mutations modify the response of poliovirus RNA to a translation initiation factor: a comparison of neurovirulent and attenuated strains. Virology 166: 394–404

Tahara SM, Morgan MA, Shatkin AJ (1981) Two forms of purified m⁷G-cap-binding protein with different effects on capped mRNA translation in extracts of uninfected and poliovirus-infected HeLa cells. J Biol Chem 256: 7691–7694

Toyoda H, Kohara M, Kataoka Y, Suganuma T, Omata T, Imura N, Nomoto A (1984) Complete nucleotide sequences of all three poliovirus serotype genomes. Implication for genetic relationship, gene function and antigenic determinants. J Mol Biol 174: 561–585

Toyoda H, Nicklin MJH, Muray MG, Anderson CW, Dunn JJ, Studier FW, Wimmer E (1986) A second virus-encoded proteinase involved in proteolytic processing of poliovirus polyprotein. Cell 45: 761–770

Trono D, Andino R, Baltimore D (1988a) An RNA sequence of hundreds of nucleotides at the 5′ end of poliovirus RNA in involved in allowing viral protein synthesis. J Virol 62: 2291–2299

Trono D, Pelletier J, Sonenberg N, Baltimore D (1988b) Translation in mammalian cells of a gene linked to the poliovirus 5′ noncoding region. Science 241: 445–448

Westrop GD, Wareham KA, Evans DMA, Dunn G, Minor PD, Magrath DI, Taffs F, Marsden S, Skinner MA, Schild GC, Almond JW (1989) Genetic basis of attenuation of the Sabin type 3 oral poliovirus vaccine. J Virol 63: 1338–1344

Yogo Y, Wimmer E (1972) Polyadenylic acid at the 3′-terminus of poliovirus RNA. Proc Natl Acad Sci USA 69: 1877–1882

Picornavirus Protein Processing—Enzymes, Substrates, and Genetic Regulation*

M. A. Lawson and B. L. Semler

1 Introduction

The sequence of events leading to the successful completion of a picornavirus infection of susceptible cells is ultimately controlled by proteolytic processing. As a consequence of encoding their viral-specific polypeptides within a single, open

Department of Microbiology and Molecular Genetics, College of Medicine, University of California, Irvine, CA 92717, USA

* Work described from the authors' laboratory was supported by a grant from the US Public Health Service (AI22693). M.A.L. is a predoctoral trainee of the US Public Health Service (CA09054). B.L.S. is the recipient of a Research Career Development Award (AI00721) from the National Institutes of Health.

reading frame, picornaviruses must depend upon the intramolecular and intermolecular interactions of viral proteinases and their cognate substrates. This chapter will first provide an overview of how the biosynthetic activities that occur during a picornavirus life cycle are regulated by protein processing activities and signals encoded in the viral genome. An examination of the nature of picornavirus proteinases and their polyprotein substrates will be presented in order to underscore the unifying principles of proteolytic cleavage and to point out the peculiar differences in processing strategies among the different picornaviruses. The application of recombinant DNA methodologies, particularly site-specific mutagenesis, to the study of structure/function relationships of picornavirus cleavage activities will then be discussed. This discussion will also focus on the molecular genetics of viable virus mutants with engineered processing lesions and on in vitro expression of altered cleavage phenotypes. Finally, the biochemical implications of the observed picornavirus processing activities will be addressed. In particular, primary sequence versus conformational determinants of protein processing will be analyzed, as well as the importance of *cis* versus *trans* cleavage. Other reviews of picornavirus protein processing have been published recently and they provide insights and details that will complement the present treatment of this topic (KRÄUSSLICH and WIMMER 1988; KRÄUSSLICH et al. 1988; JACKSON 1989; DEWALT and SEMLER 1989).

2 Regulation of Poliovirus Gene Expression by Protein Processing

2.1 Early Events During Viral Infection

The successful initiation of a picornavirus infection requires adsorption of virus particles to specific cellular receptors present on susceptible cells. Identification of such receptors for human rhinovirus (HRV) and poliovirus (PV) has recently been reported (GREVE et al. 1989; STAUNTON et al. 1989; MENDELSOHN et al. 1989). The process of penetration and uncoating releases the viral genomic RNA into the cytoplasm of the cell. Picornavirus RNAs are message-sense genomes containing a 5'-genome-linked protein (VPg) and 3'-polyadenylate tract (LEE et al. 1977; FLANEGAN et al. 1977; ARMSTRONG et al. 1972; YOGO and WIMMER 1972). Because the infecting virions contain no diffusable proteins used in viral RNA replication, the first biosynthetic phase required in the virus life cycle is protein synthesis. The products of viral-specific translation contain viral proteinases capable of intramolecular and intermolecular cleavage (HANECAK et al. 1982, 1984; TOYODA et al. 1986; PALMENBERG and RUECKERT 1982; STREBEL and BECK 1986; YPMA-WONG and SEMLER 1987a; NICKLIN et al. 1987). Collectively, there have been three different types of proteinase activities identified for the picornaviruses (L, 2A, and 3C). For the enteroviruses (poliovirus and coxsackievirus) and rhinoviruses, after ribosomes have traversed the P1-2A region of the viral genome (refer to Fig. 1), the 2A portion of this precursor is able to cleave between its own amino terminus and the carboxy terminus of VP10. It has been suggested that the P1 precursor liberated by such a 2A cleavage activity is the authentic substrate for in vitro processing by the 3C proteinase activity that produces capsid proteins (NICKLIN et al. 1987; YPMA-WONG and SEMLER 1987b). For encephalomyocardit-is virus (EMCV), foot-and-mouth disease virus (FMDV), and Theiler's murine encephalomyelitis virus (TMEV), the 2A-like cleavage most likely occurs between 2A and 2B to generate, an L-P1-2A or P1-2A precursor polypeptide (JACKSON 1986; VAKHARIA et al. 1987; CLARKE and SANGAR 1988; ROOS et al. 1989a; refer to Fig. 1). The cleavage between P1 and 2A appears to be mediated by 3C for both EMCV and FMDV (JACKSON 1986; PARKS et al. 1986; VAKHARIA et al. 1987; CLARKE and SANGAR 1988) and also for TMEV (ROOS et al. 1989a). The ribosomes then proceed through the remainder of the P2 and P3 regions of the viral genome.

Encoded in the amino-terminal portion of the P3 region is the nucleotide sequence specifying viral proteinase 3C (Fig. 1). Data from pulse-chase labeling experiments, bacterial expression of cDNA clones, and in vitro translation of

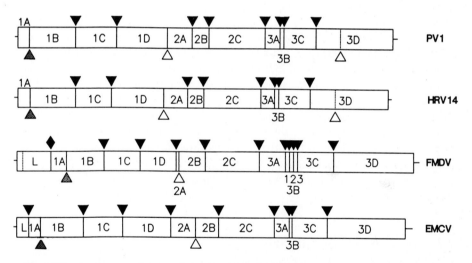

Fig. 1. The genetic maps of picornaviruses representative of the genera *Enterovirus* (*PV1*), *Rhinovirus* (*HRV14*), *Aphthovirus* (*FMDV*), and *Cardiovirus* (*EMCV*). Peptides are designated according to the standard L-4-3-4 convention (RUECKERT and WIMMER 1984). Polyprotein cleavage events mediated by the respective 3C proteinase activities are designated above the genome maps by (▼). Cleavage events known or thought to be carried out by the 2A proteinase are shown below the genetic maps by (△). Alternate products generated by 2A-mediated cleavage which do not occur in all members of the respective genera are shown by a *dotted line*. The cleavage event mediated by the L proteinase of FMDV is designated above the map by (◆). The cleavage of 1AB, which has not been assigned to any viral or host proteolytic activity, is indicated below the maps by (▲). The *dotted line* in the L protein of FMDV indicates the additional translational start site present in these virus genomes. (Adapted from ARNOLD et al. 1987)

viral RNAs suggest that the amino-terminal cleavage of 3C occurs very quickly, perhaps by an autocatalytic cleavage mechanism that proceeds as soon as the 3C protein is synthesized and can fold into an active conformation (RUECKERT et al. 1979; PALMENBERG and RUECKERT 1982; HANECAK et al. 1984; JACKSON 1986). As will be discussed in detail below, cleavage by the 3C proteinase activity may occur in *cis* or *trans* and may involve the activity of precursor polypeptides containing the 3C coding sequence. The products of translation and protein processing of P2 and P3 precursors include polypeptides that are required for RNA replication. The faithful generation of these proteins early in the infectious cycle is critical for a successful viral infection. The replication proteins must actively (and efficiently) copy some of the input viral RNAs into minus strand copies of the viral genome to initiate the process of template amplification.

The generation of replication proteins early in the picornavirus infectious cycle requires specific and efficient recognition and cleavage events. As a result, some of these cleavages (e.g., P3 → 3AB + 3CD or 3CD → 3C + 3D) can occur in *cis*, thereby bypassing the kinetic shortcomings of a bimolecular process between two polypeptides that may not be in close proximity within the cytoplasm of an infected cell (especially early in infection). Such a mechanism may also be operational in the generation of 2A cleavage activity. It has recently been shown that in vitro cleavage of the p220 component of eukaryotic initiation factor 4F (elF-4F) was induced by poliovirus protein 2A, albeit by an indirect mechanism (KRÄUSSLICH et al. 1987; LLOYD et al. 1988). The cleavage of p220 has been implicated as one of the mechanisms that enteroviruses and human rhinoviruses employ to shut off translation of capped mRNAs early after infection (ETCHISON et al. 1982, 1984; SONENBERG 1987; LLOYD et al. 1988). As mentioned above, for these viruses, protein 2A acts as the proteinase that intramolecularly cleaves the P1-P2 bond. The liberated 2A polypeptide may then go on to carry out its function(s) in host shut off. Note that the carboxy terminus of 2A contains a Q-G pair that is ultimately cleaved by 3C. KRÄUSSLICH et al. (1987) have shown that cleavage of the Q-G bond in vitro is not required for production of the p220 cleavage activity. Thus, the virus may not depend on a *trans* cleavage by the downstream 3C proteinase activity to initiate one of the processes proposed to be involved in host shut off.

2.2 Gene Product Amplification Events

Following the early rounds of viral-specific protein and RNA synthesis in picornavirus-infected cells, there is a rapid takeover by the virus of the cellular biosynthetic machinery. As a result, there is a rapid shut off of host macro-molecular synthesis and the specific accumulation of viral gene products. The proteins involved in RNA replication generate complexes with template RNAs and cellular components to produce a membrane-bound replication structure that actively synthesizes viral-specific RNAs (PENMAN et al. 1964; GIRARD et al. 1967; CALIGUIRI and TAMM 1970; LOESCH and ARLINGHAUS 1974). The generation of the enzymatic proteins that participate in viral RNA synthesis and the actual product RNAs themselves is directly regulated by 3C processing activity. All of the viral proteins that have been implicated in RNA synthesis activity (e.g., 3D, 3CD, 2C, and 3AB) are the products of specific cleavage by the 3C proteinase activity. In addition, the progeny viral RNAs all contain a VPg molecule covalently attached to their 5'-terminal uridylic acid residue by a phosphodiester bond to a tyrosine residue within the genome-linked protein (NOMOTO et al. 1977; FLANEGAN et al. 1977; ROTHBERG et al. 1978; AMBROS and BALTIMORE 1978; VARTAPETIAN et al. 1980; KING et al. 1980). Both the amino and carboxy-terminal cleavage sites that flank the amino acid sequence of VPg are cleaved by proteinase 3C (KITAMURA et al. 1981; SEMLER et al. 1981b; HANECAK et al. 1982). There have been numerous proposals based upon in vitro RNA synthesis experiments that suggest possible mechanisms for how VPg is attached to the 5' ends of progeny RNAs (DASGUPTA et al. 1980; TAKEGAMI et al. 1983; MORROW et al. 1984; YOUNG et al. 1985; ANDREWS and BALTIMORE 1986; TAKEDA et al. 1986). Independent of the precise mechanism of VPg attachment to RNA chains, the production of VPg from a precursor polypeptide requires two peptide bonds to be cleaved by 3C proteinase activity prior to or shortly after the initiation of RNA synthesis. Thus, the control of the catalytic as well as the noncatalytic polypeptides that participate in viral RNA replication is exerted at the level of protein processing by the 3C proteinase activity.

The progeny plus strand RNAs (that are the products of membrane-bound replication) are then used as templates for further protein or RNA synthesis or they associate with procapsid structures and eventually are packaged into virions. The regulatory processes that control the ultimate fate of the progeny RNAs are not well understood. Perhaps it is the relative concentration of preformed procapsids in infected cells that determines the probability of viral RNA associating with virion precursors. There may also be a contribution from a cellular enzyme that was shown to cleave VPg from the 5'-ends of poliovirus RNA to generate the mRNA (containing a 5'-pUp) found associated with polyribosomes (AMBROS et al. 1978). This enzyme (called the unlinking enzyme) is present in uninfected HeLa cells as well as in several other different mammalian cell types. It is possible that picornaviruses use the unlinking activity to regulate which progeny RNAs are destined to be mRNA and which RNAs are to be packaged into virion particles (since only VPg-linked RNA is packaged).

The increased number of picornavirus mRNAs associated with cellular polyribosomes and actively engaged in viral-specific translation ultimately yields large numbers of viral-specific proteins. Among these proteins are the 3C-containing polypeptides responsible for cleavage at specfic sites. Unlike the early phases of the infectious cycle, this product-amplification period produces sufficient quantities of 3C (or 3CD) proteinase molecules to allow *trans* cleavages of precursor polypeptides. As a result, the half-life of viral precursor polypeptides during this period is short (relative to that seen early during infection) and the protein profile on a polyacrylamide gel consists primarily of final cleavage products.

2.3 Late Events

The regulation of the late events that occur during a picornavirus infection of cultured cells is also controlled, in part, by proteolytic processing. The generation of morphogenetic precursors to progeny virion particles depends upon an ordered set of events that requires the precise folding and cleavage of the P1 precursor polypeptide (PUTNAK and PHILLIPS 1981; RUECKERT 1985; ARNOLD et al. 1987). As mentioned above, the generation of poliovirus polypeptide P1 (and by analogy, the P1 of coxsackievirus and human rhinovirus) itself requires the 2A proteinase activity. The liberated P1 polypeptide then folds into a protomer structure that sediments in a sucrose gradient as a 5S structure. Cleavage of this structure to produce VP0, VP3, and VP1 appears to be a prerequisite for the formation of the next morphogenetic intermediate, the 14S pentamer (PUTNAK and PHILLIPS 1981; RUECKERT 1985). It has been demonstrated for EMCV that site-specific mutagenesis of the VP3-VP1 Q-G site to a P-G site results in no in vitro cleavage of the site by the 3C proteinase. In addition, the lack of cleavage of the VP3-VP1 scissile bond prevents the formation of the 14S pentamer from the 5S protomer precursor (PARKS and PALMENBERG 1987). As suggested by ARNOLD et al. (1987), the release of the amino terminus of VP3 after proteolytic cleavage may generate (and stabilize) the 14S pentamers by self-association into a β-cylinder structure.

Finally, one of the last steps in virion morphogenesis is the cleavage of VP0 in procapsids to VP4 and VP2. This cleavage only occurs after procapsids have associated with a viral RNA that will be encapsidated in the mature virion particle (PUTNAK and PHILLIPS 1981; JACOBSON and BALTIMORE 1968a). The sites of proteolytic cleavage in VP0 are not among the dipeptides cleaved by the known viral proteinases. In addition, structural determinations from X-ray crystallographic data obtained for HRV14, poliovirus type 1 (PV1), and Mengo virus suggest that this cleavage could be accomplished by an intramolecular serine protease-like activity (ROSSMANN et al. 1985; HOGLE et al. 1985; LUO et al. 1987; ARNOLD et al. 1987). Such a mechanism was proposed based upon the proximity of a nucleophilic serine residue to a nearby asparagine within VP0. Following removal of a proton from the serine hydroxyl (possibly by one of the bases of the virion RNA being encapsidated), a nucleophilic attack on the peptide bond adjacent to the nearby Asn residue would result in the generation of VP4 and VP2. It should be noted that the recently solved structure of FMDV does not contain a structurally homologous region of VP0 that could accommodate such a proposed nucleophilic attack (ACHARYA et al. 1989). Although more experimental evidence is required for proof of the precise mechanism of VP0 cleavage, it is clear that protein processing is a key element in controlling the final maturation steps during the assembly of progeny virions.

3 The Nature of Picornavirus Proteinases and Their Substrates

3.1 Evidence for Polyproteins and Precursors

Two lines of investigation led to the description of proteolytic processing in the formation of picornavirus proteins: studies of the formation of the poliovirus type 1 (PV1) virion particle and studies of the synthesis of virus specific proteins in infected cells. The latter studies also led to the discovery of large polyproteins not found in the virion. Comparison of the proteins present in infectious 150S poliovirus particles containing 1A, 1B, 1C, and 1D (VP4, VP2, VP3, and VP1) with those of naturally occurring 74S empty capsids containing 1AB (VPO), 1C, and 1D and with virus particles disrupted in vitro which contained 1B, 1C, and 1D showed that the naturally occurring empty virus particles were not breakdown products of complete virions (MAIZEL et al. 1967). It was proposed by MAIZEL and colleagues that the 74S particles were possibly precursors to whole infectious virions. The relationship of the 1AB-containing 74S particle to the infectious 150S virion was more clearly delineated by JACOBSON and BALTIMORE (1968a). By using the newly described technique of inhibiting poliovirus replication in the presence of guanidine (PENMAN and SUMMERS 1965; SUMMERS et al. 1965), they were able to show the accumulation of 74S particles in infected HeLa cells. After the release from guanidine inhibition, the 74S particles were observed to form complete (infectious) 150S virions. Analysis of the composition of the two particles confirmed previous data that the 74S noninfectious particles contained 1AB and the complete virions contained 1A and 1B almost exclusively. The dynamic relationship between the 74S and the 150S particles allowed the interpretation that the 74S particle was a procapsid structure serving as a precursor to the infectious virion. The loss of 1AB concurrent with the appearance of 1A and 1B also suggested to JACOBSON and BALTIMORE (1968a) that 1AB was proteolytically cleaved to form 1A and 1B.

A more detailed description of proteolytic processing and the demonstration of large polyproteins in the infected cell resulted from attempts to account for the estimated coding capacity of poliovirus RNA. Studies of the physical and chemical properties of the poliovirion suggested to early researchers that the virus particle contained an RNA molecule of approximately 2×10^6 Da (SCHAFFER and SCHWERDT 1959; SCHAFFER 1962). A messenger RNA of that size was expected to encode a total protein mass of about 220 kDa. The values estimated were very close to the actual values of 2.4×10^6 Da and 247 kDa for the RNA genome and encoded polyprotein, respectively (GRANBOULAN and GIRARD 1969; KITAMURA et al. 1981; RACANIELLO and BALTIMORE 1981). The viral capsid proteins were the only clearly identified virus specific proteins and had, according to then current measurements, an average molecular weight of 27 kDa (MAIZEL 1963), utilizing about one-half of the expected coding capacity of the viral genome. It was expected that the known virus-directed functions found in the infected cell, the ability to replicate viral RNA (BALTIMORE et al. 1963), and the disruption of host cell protein and RNA synthesis (SALZMAN et al. 1959; ZIMMERMAN et al. 1963) would be attributed to a few as yet unidentified proteins. Analysis of the proteins synthesized in infected cells after the inhibition of host mRNA synthesis showed the presence of many more proteins in infected cells of greater apparent molecular weight than suggested by the remaining coding capacity of the genome (SUMMERS et al. 1965). Pulse-labelling of virus proteins at different times after infection also showed a shift in the proportions of protein of higher molecular weight early during infection, to lower molecular weight proteins at later times. The seemingly differential synthesis of virus proteins throughout the infection led SUMMERS and coworkers to conclude that the genomic RNA of poliovirus was multicistronic in nature. Differential expression of cistrons in a single mRNA could then be explained by a model of independent cistrons subject to differential rates of translation throughout the infectious cycle. Precedence for such a model was found in the operon theory of gene regulation in prokaryotes.

Explaining the source of the large virus specific proteins in infected cells grew more puzzling when the composition of the mRNA isolated from the polyribosomes in infected cells was found to be identical to that isolated from infectious virions (SUMMERS and LEVINTOW 1965), ruling out the possible contribution of protein by mRNA of host-cell origin, or from other forms of viral RNA. Rigorous molecular weight determinations of the virus specific proteins along with pulse chase experiments of the kinetics of viral protein synthesis in infected cells helped account for the apparent discrepancy in protein mass versus coding capacity (SUMMERS and MAIZEL 1968; MAIZEL and SUMMERS 1968). Although the combined mass of all virus specific proteins was approximately 500 kDa, several of the proteins could be demonstrated to be primary translation products, while others were secondary products that coincidentally appeared as the amount of primary products decreased. It was concluded by SUMMERS and MAIZEL (1968) that the larger molecular weight proteins were precursors to the smaller capsid proteins and that the smaller proteins were most likely produced by specific proteolytic events. They also proposed that proteolysis would serve to regulate the expression of a particular protein function. The processing model of gene expression developed by this group did not directly argue against a multicistronic genomic RNA, but did reconcile the apparent difference between the number of virus proteins and the coding capacity of the genome. The processing of large polypeptides to smaller proteins during PV1 infection was confirmed and expanded to other picornaviruses by HOLLAND and KIEHN (1968). Using the enteroviruses poliovirus types 1 and 2, coxsackievirus B1 and B5, and the cardiovirus Mengo virus they showed that the processing of viral proteins in infected cells was not dependent on the host cell but rather determined by the infecting virus. HOLLAND and KIEHN (1968) also found by following virus particle formation that there were distinct primary and secondary cleavage events that led to the production of a mature virus particle.

A complete model describing the role of proteolytic processing in poliovirus gene expression was presented by JACOBSON and BALTIMORE (1968b). By employing amino acid analogues to partially inhibit protein processing in infected cells, they detected a large polypeptide, NCVP0 (P2-P3), the presence of which coincided with the lack of the smaller P2 and P3 proteins. An even larger protein, NCVP00 (P1-P2-P3) was also detected. The larger precursor proteins were better demonstrated in PV1-infected cells by the use of nonspecific protease inhibitors (JACOBSON et al. 1970). They were also more easily detected in coxsackievirus B1 infected cells under normal conditions (KIEHN and HOLLAND 1970) and in cells infected with PV1 at elevated temperatures (BALTIMORE 1971; GARFINKLE and TERSHAK 1971). It was also noted that under conditions which severely inhibited protein processing, proteins smaller than P2 were not produced. The above data provided evidence for a proteolytic cascade in which a large polyprotein was synthesized and subsequently cleaved to form the primary proteins P1, P2, and P3. These three proteins would then be individually processed to form the other proteins involved the virus life cycle. Not only did this model refine the conclusions by SUMMERS and MAIZEL (1968), but also suggested to JACOBSON and BALTIMORE (1968b) that, in general, a eukaryotic messenger RNA must be monocistronic in nature, that is, bear only one utilized start codon and one utilized stop codon. The somewhat peculiar protein processing method of gene expression employed by picornaviruses may largely be the way this group of viruses deals with the requirement for monocistronic mRNA dictated by the eukaryotic translational machinery. In a broader view, the methods of gene expression used by other RNA viruses of the positive, negative, and double-stranded variety may also be the result of adaptation to this requirement.

3.2 Gene Order Determination

The hypothesis of a single open reading frame encoding all the picornavirus specific proteins was upheld by experiments designed to determine the order in which the proteins occurred in the genome. Should translation be initiated at a discrete location on the genomic RNA, the order of occurrence of the various gene products could be determined by examination of the kinetics of virus specific protein synthesis or of the incorporation of radiolabeled amino acids into virus specific protein after inhibition of the initiation of translation. Experiments which used the latter approach took advantage of the antibiotic pactamycin, which at low concentrations inhibits translation by preferentially blocking initiation rather than elongation, allowing only previously initiated peptides to be completed. The addition of radioactive amino acids after or concurrent with the addition of the drug would show, upon polyacrylamide gel electrophoresis, a quantifiable decrease in the amount of radiolabeled amino acids incorporated into the completed proteins, the amino terminal sequences being the most affected and incorporating the least amount of label. Such experiments (TABER et al. 1971; SUMMERS and MAIZEL 1971) ordered the primary gene products of poliovirus as P1a, P2x, and P1b (P1, 2C, and P3). A finer analysis by REKOSH ordered the capsid proteins in the P1 precursor as VP4, VP2, VP3, and VP1 (1A, 1B, 1C, and 1D; REKOSH 1972).

Similar gene orders were determined for the gene products of the cardiovirus EMCV, human rhinovirus type 1 (HRV1), and Mengo virus, and the same relative amounts of analogous peptides were synthesized by several picorna-viruses (BUTTERWORTH and RUECKERT 1972a; MCCLEAN and RUECKERT 1973; PAUCHA et al. 1974; BUTTERWORTH and KORANT 1974) indicating that picorna-viruses utilized the same general scheme of translation and processing (BUTTERWORTH 1973). Detailed kinetic analysis of proteins synthesized in EMCV-infected HeLa cells further improved the genetic map of the genome and ordered much of the EMCV proteolytic cascade (BUTTERWORTH et al. 1971; BUTTERWORTH and RUECKERT 1972b). The relationship of the various picorna-viral proteins to each other as well as their map locations were upheld by the more biochemical methodology of analysis by tryptic and CNBr digests of partially purified proteins (JACOBSON et al. 1970; BUTTERWORTH et al. 1971; ABRAHAM and COOPER 1975; SANGAR et al. 1977; DOEL et al. 1978; RUECKERT et al. 1979; SVITKIN et al. 1979; KEW et al. 1980; WIEGERS and DERNICK 1981a, b; SEMLER et al. 1982; GRUBMAN et al. 1984; PALLANSCH et al. 1984). The validity of the gene order determinations for poliovirus was confirmed by sequencing the viral genome (KITAMURA et al. 1981; RACANIELLO and BALTIMORE 1981) and precisely determining the amino acid sequences cleaved in the various processing events (EMINI et al. 1982; SEMLER et al. 1981a, b; LARSEN et al. 1982; ADLER et al. 1983). The genetic maps of other picornaviruses are derived from the cloned genomic RNAs and by comparison with each other and their known protein products (BOOTHROYD et al. 1982; STANWAY et al. 1984a, b; CALLAHAN et al. 1985; CARROLL et al. 1984; PALMENBERG et al. 1984; FORSS et al. 1984; ROBERTSON et al. 1985; LINDBERG et al. 1987; IIZUKA et al. 1987; JENKINS et al. 1987; TRACY et al. 1985; LIPTON et al. 1984; PEVEAR et al. 1987; OHARA et al. 1988; TICEHURST et al. 1989; ROOS et al. 1989a). The currently accepted genomic maps of the several genera of picornaviruses are shown in Fig. 1.

3.3 Evidence for and Identification of Viral Proteinases

3.3.1 Evidence for Virus Encoded Proteinases

That the proteolysis of picornaviral proteins was specific to viral proteins and may be a virus-specified function was noted by KIEHN and HOLLAND (1970) who observed that the proteolytic activity increased over time during infection. Inhibition of proteolysis through the use of amino acid analogues did not influence the overall size distribution of HeLa cell proteins as it did the virus specific proteins (individual cellular proteins could not be followed by this analysis). The explanation offered for this phenomenon was that the proteolytic activity involved in processing viral proteins specifically was either a cellular protein activated by infection or a proteinase encoded by the virus itself. It was well established that the cleavage of the poliovirus and EMCV capsid precursor occurred while the polyprotein was *in statu nascendi* (JACOBSON et al. 1970; BUTTERWORTH and RUECKERT 1972b). Due to the efficiency of these cleavage events, the identification of the activity responsible for their mediation as host or virus specified was difficult. Several lines of experimental evidence suggested that at least an initial cleavage event was mediated by the host cell during poliovirus and FMDV infection (KORANT 1972; BURROUGHS et al. 1984) and during translation of poliovirus RNA in an in vitro system (VILLA-KOMAROFF et al. 1975).

3.3.2 Identification of 3C As a Virus Specific Proteinase

Direct evidence for a virus encodd proteinase was provided by LAWRENCE and THACH (1975) who clearly correlated the increase in EMCV-specific proteolytic activity with increased viral protein synthesis. Partial purification of the proteinase activity was also attained, and it was thought to be associated with the capsid protein 1C (γ). An equivalent purification of a poliovirus-specific enzyme was attained by KORANT et al. (1979) who tentatively identified the proteinase as 2C (2x). The use of a rabbit reticulocyte lysate in vitro translation system (PELHAM and JACKSON 1976) in combination with the translation inhibitors edeine and emetine allowed Pelham to demonstrate that the enzyme responsible for the majority of EMCV processing was, in fact, virus encoded and present in a protein other than the capsid precursor (PELHAM 1978). Kinetic studies of the synthesis of EMCV proteins in vitro suggested to Shih et al. (1979) that the EMCV viral proteinase was a 22 kDa virus-specific peptide (3C) rather than the 23 kDa peptide 1C. Biochemical evidence supporting this was presented by Gorbalenya et al. (1979) who purified the activity of the EMCV proteinase as a 22 kDa virus specified protein. The 22 kDa peptide was purified from in vitro translation reactions and mapped to the 3′ end of the viral genome (SVITKIN et al. 1979; PALMENBERG et al. 1979). Clever experimentation gave clear evidence that the 22 kDa 3C protein of EMCV was itself a proteinase rather than an activator of a host cell proteolytic activity (GORBALENYA et al. 1981). In their discussion these workers proposed by argument of analogous genome structure that the correct poliovirus proteinase was the 3C protein, as in EMCV. The validity of such an argument was affirmed by experiments (HANECAK et al. 1982) in which antibodies to poliovirus 3C specifically inhibited proteolysis at Q-G amino acid cleavage sites of proteins synthesized in a HeLa cell in vitro translation system (CELMA and EHRENFELD 1975), whereas 2C-specific antibodies did not have an effect on any cleavage event. It was noted then that the rapid cleavage event that occurred at a Y-G sequence at the P1-P2 junction as well as the cleavage at the 3C′/3D′ junction were not affected, leaving open the possibility of host cell proteases at work. The 3C proteinases of PV1 and PV2 as well as that of HRV14 have since been expressed in *E. coli* and their activities as proteinases have been characterized (HANECAK et al. 1984; IVANOFF et al. 1986; NICKLIN et al. 1988; LIBBY et al. 1988; PALLAI et al. 1989). The PV2 and the HRV14 enzymes have been purified to apparent homogeneity (NICKLIN et al. 1988; LIBBY et al. 1988). The major proteolytic activity of the aphthovirus FMDV was mapped to the 3' end of the virus genome in a manner similar to that of PELHAM, above (GRUBMAN and BAXT 1982). The data assigning the source of proteolytic activity of FMDV to the 3C region of the genome was presented by KLUMP et al. (1984). However, rigorous determination of 3C as an authentic proteinase rather than an activator of proteolytic activity was not carried out. Recent experiments involving the cardiovirus TMEV by ROOS and coworkers (1989a) have shown that the majority of cleavage events in TMEV protein processing are mediated by the 3C protein. The above experiments clearly delineated the active proteolytic role of 3C in EMCV, PV, FMDV, and TMEV protein processing. The central role of 3C-mediated processing may be considered a general characteristic of picornavirus gene expression.

3.3.3 Identification of 2A as a Virus Specific Proteinase

Identification of the enzyme responsible for rapid removal of the capsid precursor from the nascent polypeptide was elusive and required a more easily manipulated system for analyzing picornavirus directed protein synthesis than that allowed by analysis of infected cells or crude cell extracts. Consequently, the identification of additional viral proteinases was almost entirely dependent on the availability of cloned cDNAs of picornavirus genomes and their use in in vitro transcription and translation systems.

The strongest evidence for a second proteinase encoded by a picornavirus was the identification of the poliovirus 2A protein as a Y-G specific proteinase responsible for the rapid removal of the capsid precursor P1 from the nascent polyprotein (TOYODA et al. 1986). As in the identification of the poliovirus 3C proteinase by HANECAK et al. (1982) described above, several experimental approaches were taken to show the proteolytic activity of 2A. Expression of wild type and mutant poliovirus 2A cDNAs in E. coli and in infected HeLa cell extracts coupled with antibody inhibition experiments were employed to designate the primary structure of the 2A protein as the entity responsible for cleavage at Y-G sequences in the poliovirus polyprotein. The conclusions of TOYODA et al. were supported by further in vitro experiments involving translation in a rabbit reticulocyte lysate system of in vitro synthesized subgenomic RNAs (NICKLIN et al. 1987). The poliovirus 2A proteinase has been purified from infected HeLa cells to near homogeneity and the activity of the enzyme using synthetic substrates has been characterized (KÖNIG and ROSENWIRTH 1988).

Aside from its direct role in the initial cleavage of the nascent polyprotein, the 2A proteinase of enteroviruses and of rhinoviruses has been shown to be partially responsible for host shut-off by causing indirectly the cleavage of the p220 component of eIF-4F. The degradation of p220 has been shown to limit the ability of the host cell translational apparatus to utilize capped mRNAs, thereby enhancing the translation of the uncapped viral RNA (ETCHISON et al. 1982, 1984; BONNEAU and SONENBERG 1987; LLOYD et al. 1986, 1987, 1988; KRÄUSSLICH et al.

1987). The exact mechanism by which the cleavage of p220 is directed is still obscure and requires further mutational and biochemical analysis.

In an in vitro study that exploited the efficient translation of EMCV viral RNA in the rabbit reticulocyte lysate, JACKSON (1986) was able to show in a detailed kinetic analysis that the resistance to protease inhibitors and the extremely efficient nature of the LP12A/2B cleavage event (the cardiovirus and aphthovirus homologue of the P1/2A cleavage of enteroviruses and rhinoviruses) could be best explained by an intramolecular cleavage mechanism. The evidence presented also argued for a proteolytic enzyme present in the 5′ half of the genome (JACKSON 1986). Support for this came from in vitro expression of RNA synthesized from truncated subgenomic cDNA clones containing only the LP12AB regions of the genome (PARKS et al. 1986). The expression of such subgenomic RNAs in rabbit reticulocyte lysates generated authentic LP12A substrates for cleavage by cotranslated or posttranslationally added EMCV 3C proteinase. An accurate determination of the enzyme responsible for the LP12A/2B cleavage in EMCV has yet to be made. Recent experiments with the aphthovirus FMDV (VAKHARIA et al. 1987; CLARKE and SANGER 1988) and with the cardiovirus TMEV (ROOS et al. 1989a) in in vitro translation systems have also implicated the 2A protein in a proteolytic role, perhaps via an intramolecular reaction. The predicted amino acid sequences of the 2A proteins of EMCV, TMEV, and FMDV do not contain sequences thought to be required for the formation of a catalytic center. Interestingly, the 2A region of FMDV contains only 16 amino acids. The cleavage of the LP12A/2B junction in the polyproteins of FMDV and TMEV does not require the presence of functional 3C or L proteins (CLARKE and SANGER 1988; ROOS et al. 1989a) but clear determination of the responsible protein has not been forthcoming. An unpopular but legitimate possibility still exists that a host cell enzyme is responsible for this event.

3.3.4 The Proteolytic Activity of the FMDV L Protein

Building upon the work of BURROUGHS et al. (1984) which showed that more than one proteinase was involved in the processing of the FMDV polyprotein, STREBEL and BECK (1986) demonstrated by mutational analysis of LP1 subgenomic cDNAs and subsequent expression in E. coli that the L protein was responsible for mediating the LP/1 cleavage. Expression in vitro of RNA transcribed from wild type and mutant cDNAs bearing lesions in the L coding region showed that the L/P1 cleavage was sensitive to L mutation or deletion, and to antibodies directed against the L protein, whereas similar treatment of the P1 sequences did not have an effect. The L protein is found in as many as three forms in infected cells and under some in vitro translation conditions. Two of these are attributed to the alternate use of initiating AUG codons in the FMDV RNA and result in an L protein of either 20 kDa (La) or 16 kDa (Lb; CLARKE et al. 1985; SANGER et al. 1987). The third form of L, Lb′, is derived from Lb, possibly by the removal of carboxy-terminal amino acids via some unidentified processing mechanism

(SANGAR et al. 1988). Whether the alternate forms of L are functionally different is not known.

There are unique differences between the L proteins of FMDV and of the cardioviruses EMCV and TMEV. Although both aphthoviruses and cardioviruses possess L proteins, only those of FMDV possess proteolytic activity. The FMDV L proteins catalyze the cleavage of the L/P1 bond while the cleavage of the L/P1 junction in EMCV is carried out by 3C (PARKS et al. 1986). Experiments by ROOS et al. (1989a) indicate that like EMCV, the TMEV L protein does not possess proteolytic activity. Due to the sequence similarity between EMCV and TMEV it is likely that the TMEV L protein is also removed from the LP12A precursor by the 3C proteinase (OHARA et al. 1988; ROOS et al. 1989a). Another significant difference between the cardio- and aphthovirus L proteins is the role of L in host cell shut-off. The FMDV L protein has recently been shown to mediate the cleavage of p220 as does the poliovirus and rhinovirus 2A proteins (DEVANEY et al. 1988; LLOYD et al. 1988). It is tempting to speculate that the ability to rapidly inhibit the translation of capped mRNA during infection is related to the proteolytic activities of the L and 2A proteins, and that the cardioviruses have somehow lost or never evolved the ability to inhibit host cell translation in this way (MOSENSKIS et al. 1985).

3.3.5 Other Forms of Viral Proteinases

Refinement of in vitro translation systems (DORNER et al. 1984; YPMA-WONG and SEMLER 1987a) and more sophisticated mutational analysis of the picornavirus proteinases has shown that the various precursors of the 3C proteinase possess proteolytic activity as well (YPMA-WONG and SEMLER 1987a; JORE et al. 1988; YPMA-WONG et al. 1988b; PARKS et al. 1989). The active precursor proteins may be directly involved in regulation of gene expression by their differential activity and by modified substrate interaction. These aspects of picornavirus protein processing will be treated in more detail below.

3.4 The Nature of Picornaviral Proteinases

Placing the proteinases of picornaviruses and other related viral proteinases in one of the classical divisions of known proteases (acid proteases, metalloproteases, thiol proteases, and serine proteases) has been a task of considerable difficulty. A tentative identification of the major proteolytic (3C) activity of poliovirus as one of the trypsin-like serine protease class of enzymes was based on the inhibition of cleavage by chloromethyl ketone substrate analogues (SUMMERS et al. 1972), which act by alkylation of the imidazole ring of the active site histidine residue (CRAIK et al. 1987). The conclusions of the study by SUMMERS and coworkers were tempered by the known susceptibility of thiol proteases to the alkylating activity of the inhibiting compounds employed. Host cell effects on the sensitivity of processing to different chloromethyl ketone inhibitors was reported as well (KORANT 1972). To further cloud the issue, proteolytic processing was found to be sensitive to the thiol protease inhibitors iodoacetamide, N-ethylmaleimide, and cystatin in in vitro processing studies. The 3C proteinase of EMCV was also found to be largely sensitive to thiol protease inhibitors (KORANT 1973; KORANT et al. 1985; PELHAM 1978). More detailed studies of the purified EMCV 3C proteinase showed clearly that cysteine residues were important for catalytic activity and that the nonspecific alkylation of substrates by chloromethyl ketone serine protease inhibitors was indeed interfering with classification of the enzyme (GORBALENYA and SVITKIN 1983). Though structurally quite different, serine and thiol proteases, use similar mechanisms of catalysis, and the two classes of enzymes are taken to be an example of convergent evolution. Both classes of enzymes rely on the nucleophilic attack of the carbonyl carbon atom by an ionized serine hydroxyl or cysteine thiol group, catalyzed by a histidine residue. The serine or cysteine and histidine residues are part of a catalytic triad which additionally includes an aspartic acid residue, all of which are arranged in the active site to create an extended, hydrogen bonded system with the substrate during the formation of a stable transition state (for reviews see NEURATH 1984; POLGÁR and HALÁSZ 1982). However, the serine and cysteine residues are not interchangeable between the enzyme classes (HIGAKI et al. 1987). The role of the aspartic acid residue in peptide bond cleavage by thiol proteases is not entirely clear (POLGÁR and HALÁSZ 1982). The apparent conservation of the aspartic acid among serine, thiol, and picornaviral proteolytic enzymes does imply an integral role in the catalytic mechanism.

The sequence comparison of several picornaviruses and plant viruses revealed significant regions of similarity between them. These studies suggested that the C_{147} and H_{161} residues of the poliovirus 3C proteinase and the equivalent residues in other picornaviruses and plant viruses may serve as the catalytic residues in a proteinase active site. No conserved aspartic acid residue was found by the comparative methods used (ARGOS et al. 1984; FRANSSEN et al. 1984; WERNER et al. 1986; DOMIER et al. 1987; GORBALENYA et al. 1988). The involvement of C_{147} in catalysis has been supported by mutational analysis of poliovirus 3C proteinase subclones expressed E. coli in which the C_{147} residue was converted to a S, and the mutated protein subsequently lost all catalytic activity, as monitored by the cleavage of the 3C/3D bond. Conversion of C_{153} (a nonconserved residue) to a S had only a negligible effect. Support for the involvement of H_{161} was also reported, as conversion to G also abolished proteolytic activity (IVANOFF et al. 1986). A more recent amino acid sequence comparison biased toward predicted secondary structure of the picornavirus 3C and 2A proteinases rearranged the proposed active site residues to include H_{40}, D_{85}, and C_{147} for the 3C proteinase and H_{20}, D_{38}, and C_{109} for the 2A proteinase. The study also showed that the 3C and 2A proteinases resemble the large and small trypsin-like serine proteases, respectively, in structure, but suggested that in the mechanism of catalysis, they may resemble that of the cysteine-utilizing thiol proteases (BAZAN and FLETTERICK 1988).

Other studies of 3C proteinase similarity to cellular enzymes have shown a relatedness to cellular thiol proteinases that extends into the 3D coding sequence (GORBALENYA et al. 1986). The dilemma of classification of the picornavirus proteinases has been somewhat resolved by the recent proposal of a superfamily of serine-like proteases that include both the cellular serine proteases and the picornavirus proteinases (GORBALENYA et al. 1989).

In summary, the picornavirus proteinases represent a unique class of enzymes that integrate the characteristics of both serine and thiol proteases. The viral proteinases should provide an indispensable tool for understanding the mechanism of proteolysis by both classes of enzymes. The picornavirus proteinases remain a target for antiviral chemotherapy, and the unique nature of the enzymes may provide an avenue for their selective inhibition. A better understanding of the structure and catalysis of picornaviral proteinases awaits the availability of a crystal structure.

3.5 The Nature of the Substrates

The picornaviral proteinases are significantly different from other proteolytic enzymes not only in their unique combination of structure and catalytic mechanism, but also in the substrates utilized. The proteinases of picornaviruses appear to have only one general role in the virus life cycle: to regulate viral gene expression through specific processing of the encoded polyprotein. Because the substrates are apparently limited and the proteinases have coevolved with them, it is reasonable to suppose that the substrates have as great a role in processing as the proteinases themselves. This cooperation is reflected in the primary amino acid sequence of the cleavage sites and in the secondary and tertiary structure of the polyprotein.

Evidence that the structure of the substrate is important in processing can be obtained by reevaluating the results of early processing studies in which poliovirus-infected cells were incubated in the presence of amino acid analogues to allow the detection of higher molecular weight precursor proteins (KIEHN and HOLLAND 1970; BALTIMORE et al. 1969; JACOBSON et al. 1970). Such treatment disrupts processing, presumably by altering the structure of proteins into which the amino acid analogues are incorporated. Because the target size of the proteinase is much smaller than that of the remaining proteins, the greater effect would be on the substrates. Misfolding of substrate precursor proteins may render them incompetent as such. Further evidence for the importance of structure in processing comes from temperature shift experiments. The processing of virus proteins was inhibited when infection was carried out at elevated temperatures and could not be reversed when returned to normally permissive temperatures (GARFINKLE and TERSHAK 1971; BALTIMORE 1971).

Insight into the different dependence on higher order structure for substrate utilization by the poliovirus 2A and 3C proteinases can also be derived from the above experiments. In retrospect, the processing events most affected by incorporation of amino acid analogues and by elevated temperature were those we now know to be 3C-mediated. The removal of the P1 capsid precursor by 2A was the least affected, indicating that 2A may be less dependent on substrate conformation for utilization than 3C. Such an interpretation is further supported by recent in vitro translation experiments which show that the processing of the poliovirus P1/2A junction is dependent on minimal sequences around the cleavage site (YPMA-WONG and SEMLER 1987b). The 3C proteinase has been purified from *E. coli* expressing the protein (NICKLIN et al. 1988; PALLAI et al. 1989). The activity of the purified protein is less efficient when tested against synthetic peptides that mimic an authentic cleavage site compared to cleavage of authentic precursor polypeptides, suggesting further that extended structural interactions are important for efficient substrate recognition and utilization.

The cleavage activity of 3C is responsible for the majority of processing events in all known picornaviruses. The processing of the capsid precursor P1 to 1AB, 1C, and 1D, as well as all P2 and P3 processing is dependent on 3C activity. In poliovirus, 3C-mediated cleavages occur at Q-G sequences exclusively, although it appears unique among picornaviruses in this characteristic (KITAMURA et al. 1981; HANECAK et al. 1982). Other enteroviruses as well as rhinoviruses, cardioviruses, and aphthoviruses can utilize other dipeptide sequences, although the variability is limited and may be governed by interaction with the active site of the proteinase (see Table 1). Not all sequences recognized by the 3C proteinases are used as cleavage sites in the polyprotein. For poliovirus, only 9 of 13 predicted Q-G pairs in the polyprotein are cleaved (the ninth being a rare event in the 3CD protein; SEMLER et al. 1983). There are additional contributions to cleavage site recognition by the primary amino acid sequence. The utilized cleavage sites of the 3C proteinases of poliovirus include an aliphatic residue in the P_4 position (where Pn denotes the *n*th residue amino terminal to the cleaved peptide bond and Pn′ denotes the *n*th residue carboxyterminal to the cleaved bond; NICKLIN et al. 1986). The amino acid residues around the cleavage sites may play some role in site recognition, but seem to be secondary to structural considerations (YPMA-WONG et al. 1988a). That the viral proteinases do not cleave at every occurrence of the recognized sequence(s) is not surprising upon consideration that the viral proteinases are not general degradative enzymes but act on specific proteins in a manner that maintains their functional integrity. The uncleaved dipeptide sequences recognized by the proteinases may also remain inaccessible to the proteinase as is seen in the capsid protein 1B of poliovirus (HOGLE et al. 1985) in which a Q-G dipeptide pair is confined within a β-barrel structure.

The 2A proteinase of enteroviruses and rhinoviruses is responsible for the separation of the capsid precursor from the polyprotein during translation of the viral RNA. Although it is widely accepted that the P1/2A cleavage occurs while the polyprotein is nascent, immediately after the synthesis of 2A, this has only been clearly demonstrated for the LP12A/2B cleavage in EMCV (JACKSON 1986). However, analysis of polyribosomes from poliovirus infected HeLa cells revealed no nascent polypeptides greater than 130 kD (JACOBSON et al. 1970). The target sequence of the 2A proteinase of enteroviruses and rhinoviruses is

Table 1. Summary of known (*italics*) and predicted cleaved amino acid pairs in several picornavirus genomes. Amino acid pairs are generally derived by alignment of sequences with those of viruses for which the cleavage sites are known (BACHRACH et al. 1973; BURRELL and COOPER 1973; LARSEN et al. 1982; PALLANSCH et al. 1984; ROBERTSON et al. 1985; SKERN et al. 1985; ZIOLA and SCRABA 1976). Question marks designate uncertainty in the potential site of cleavage predicted from a single cDNA clone. Multiple listings designate predicted sites found in separate cDNA clones. Table adapted from LINDBERG et al. 1987

Boundary	PV1-3	CVB1	CVB3	CVB4	HRV2	HRV14	EMCV	TMEV	FMDV O_1K	FMDV A_{10}	FMDV A_{12}
L/1A	NA	NA	NA	NA	NA	NA	Q/G	Q/G	G/N	G/Q G/N	G/Q? Q/S? Q/N?
1A/1B	N/S	N/S	N/S	N/S	N/S	N/S	A/D	L/D	A/D	A/D	A/D
1B/1C	Q/G	Q/G	Q/G	Q/G	Q/G	Q/G	Q/S	Q/S	Q/G	Q/G	V/G
1C/1D	Q/G	Q/G	Q/G	Q/G	Q/N	E/G? Q/T?	Q/G	Q/G	Q/T	Q/T	Q/T
1D/2A	Y/G	T/G	Q/S? I/R? Y/R? Q/N?	Y/G	Y/V?	Y/G?	E/S	E/N	L/N	L/N	L/N
2A/2B	Q/G	Q/G	Q/G	Q/G	Q/G	Q/G	Q/G	Q/G	G/P	G/P	R/P
2B/2C	Q/G	Q/N	Q/N? Y/G?	Q/N	E/S	Q/A	Q/S	Q/G	Q/L	Q/L	Q/L
2C/3A	Q/G	Q/G	Q/G	Q/G	Q/G	Q/G	Q/G	Q/S	Q/I	Q/I	Q/I
3A/3B(1)	Q/G	Q/G	Q/G	Q/G	Q/G	Q/G	Q/G	Q/A	Q/G	Q/G	Q/G
3B1/3B2	NA	NA	NA	NA	NA	NA	NA	NA	Q/G	Q/G	Q/G
3B2/3B3	NA	NA	NA	NA	NA	NA	NA	NA	Q/G	Q/G	Q/G
3B(3)/3C	Q/G	Q/G	Q/G	Q/G	Q/G	Q/G	Q/G	Q/G	Q/S	Q/S	Q/S
3C/3D	Q/G	Q/G	Q/G	Q/G	Q/G	Q/G	Q/G	Q/G	Q/G	Q/G	Q/G

different than that of 3C, generally Y-G (see Table 1). In some poliovirus and rhinovirus serotypes there is an additional cleavage site recognized by 2A in the P3 region which generates, when cleaved, the 3C′ and 3D′ proteins. This additional cleavage is not important for virus growth in cell clulture, and the function of 3C′ and 3D′ remains unclear (LEE and WIMMER 1988; see below).

An outstanding feature of the 2A proteinases of entero- and rhinoviruses, and that of the leader protein FMDV, is their role in mediating the inhibiton of capped mRNA translation during infection (KRÄUSSLICH et al. 1987; LLOYD et al. 1988; DEVANEY et al. 1988). There is convincing evidence that cleavage of the p220 component of eIF-4F is involved in the inhibition of translation of capped mRNAs (ETCHISON et al. 1982; ETCHISON and FOUT 1985) and that p220 is not directly cleaved by either the 3C or the 2A protein (LLOYD et al. 1985, 1986; LEE et al. 1985). Because the cleavage of p220 is correlated with proteolytic activity (LLOYD et al. 1988), it is possible that cleavage is the ultimate result of a proteolytic cascade initiated by the 2A or L proteins.

4 Molecular Genetics of Proteolytic Cleavage

4.1 Lesions That Affect Enzymatic Activity of the Viral Proteinases

The techniques of gene manipulation have proved invaluable to the analysis of protein processing in picornaviruses, and we have obtained a great deal of information on how picornaviruses may regulate gene expression through protein processing. Presented here briefly are mutational analyses of picorna-virus processing categorized by the enzymatic activity chiefly targeted or affected. A particular mutation in a viral proteinase will exhibit a wide variety of effects on processing or other viral functions, and often the effects seen are particular to the system in which the mutant is analyzed. The pleiotropic effects of mutations in picornavirus genes reflect the highly integrated nature of the various gene functions and, although they may make particular phenotypes difficult to understand, will eventually provide a better understanding of the regulation of the virus life cycle.

4.1.1 Lesions That Affect 3C Activity

The simplest manipulation of a cloned gene is deletion of all or part of the coding sequences and then analyzing the effect of such a drastic alteration. This approach was taken to identify the 3C region of the O_1K strain of FMDV as a proteinase (KLUMP et al. 1984). The effect of a carboxy-terminal 31 amino acid deletion of 3C was analyzed by expression in *E. coli* of subgenomic cDNA clones containing sequences from 1D through the 3C proteinase. The results showed such a deletion to cause the loss of processing activity as measured by the presence or absence of 1D. Interestingly, no processing of the 2A/2B junction was detected in these studies although it has been shown more recently that processing of this site is independent of active L and 3C proteins (VAKHARIA et al. 1987).

Bacterially expressed poliovirus 3C has been analyzed by linker insertion and site directed mutagenesis. HANECAK et al. (1984) introduced four amino acids into 3C via insertion of an octameric Sac I linker at the Bgl II site in the cDNA clone. Expression of the resulting mutant 3C proteinase in *E. coli* showed loss of 3C activity as judged by cleavage of the amino and carboxyl terminal Q-G sites of the 3C protein itself. The same insertion mutant has been analyzed in an vitro translation system programmed with RNA transcribed from a full length cDNA bearing the lesion (YPMA-WONG and SEMLER 1987a). All of the potential cleavage sites are theoretically tested by this method. In the in vitro translation system the mutated 3C was incapable of proper processing activity. Unusual proteins were seen upon translation, however their identities were not clearly established. Thus, abolition of all 3C activity could not be positively concluded. A similar insertion of four amino acids into the EMCV 3C coding sequence near the putative active site C_{159} also disrupted 3C-mediated processing. The EMCV insertion mutant was analyzed by in vitro transcription of a subgenomic cDNA clone bearing the lesion and subsequent in vitro translation of the RNA in a rabbit reticulocyte lysate (PARKS et al. 1986). The processing of only the L/P1AB and the P3 proteins were tested in this study. A duplication of the P_2 serine residue of the PV1 3C/D cleavage site disrupted processing at that site when analyzed in a bacterial expression system (SEMLER et al. 1987). Translation in vitro of the same mutation did not reveal any gross processing defects (YPMA-WONG et al. 1988b). No viable virus has been recovered by DNA transfection of the mutant cDNA (SEMLER et al. 1987).

It was shown in both in vitro translation studies above that the proteins produced by mutant cDNA clones were indeed capable of functioning as substrates. This is especially revealing in the EMCV study in which the mutant P3 protein produced in vitro was fully competent as a substrate for post translationally added wild type 3C activity from EMCV-infected cells. Wild type poliovirus P3 proteins produced in the in vitro translation system are not processed to a great extent by proteolytic activity from infected cells so the same phenomenon could not be examined. The data from the EMCV experiment can be interpreted to mean that only the activity of the proteinase was affected by the insertion. The recognition of mutant 3C as a substrate by wild type 3C activity suggests that recognition may be limited to the region of the cleavage site.

Recently the proteolytic activity of HRV14 and CVB3 was shown to function on poliovirus substrates when provided in the context of a chimeric polyprotein (DEWALT et al. 1989). Insertion of four amino acids into the CVB3 3C proteinase or deletion of part of the HRV14 3C proteinase disrupted processing of the chimeric polyprotein. Further experiments involving more exact replacement of the poliovirus 3C with that of CVB3 have shown that the amino acid insertion used above only partially disrupts processing. Additional insertion mutagenesis of the latter chimeric cDNA has also shown that the degree to which processing is disrupted is dependent on the site of the insertion. Suballelic replacement of regions of the poliovirus 3C protein with equivalent regions of the CVB3 3C protein has also indicated a differential ability of 3C to recognize structural and non structural proteins as substrates (M. A. LAWSON; B. DASMAHAPATRA and B. L. SEMLER, submitted for publication).

The tentative identification of the C_{147} and H_{161} residues as the active C-H pair in the poliovirus 3C proteinase was made by conversion of the C and H residues to S and G, respectively (IVANOFF et al. 1986). The loss of proteolytic activity of 3C due to the mutations was observed in a bacterial expression system. Conversion of the nearby C_{153} to S had a negligible effect. The only substrate tested in the study was the 3C/3D site, an inefficiently cleaved site both in vivo and in vitro. Although the mutations made were not exhaustive, the conclusion that C_{147} may be a part of the catalytic site has been supported by sequence comparison analyses (ARGOS et al. 1984; BAZAN and FLETTERICK 1988; WERNER et al. 1986).

Site-directed mutagenesis of the poliovirus 3C cDNA has proved an invaluable tool for analysis of the 3C proteinase, not only as a means of creating specific mutants, but also as a tool for creation of mutagenic cassettes to allow more thorough mutational analysis of small, defined regions of the protein. Using the mutagenic cassette methodology, DEWALT and SEMLER (1987) have constructed a series of single and double amino acid conversion mutations in the region of $aa_{51} \rightarrow aa_{61}$ of the poliovirus 3C proteinase. Viable mutant viruses were recovered by DNA transfection of COS-1 monolayers. Upon analysis, the mutant viruses were found to have the general defect of altered recognition of the 3C/3D cleavage site, producing either excess or depressed amounts of 3C and 3D proteins (DEWALT and SEMLER 1987; DEWALT et al. 1990). A broad conclusion of these studies was that the mutagenized region, a proposed loop between β-barrel secondary structures (WERNER et al. 1986; BAZAN and FLETTERICK 1988), was involved in efficient recognition of cleavage of the 3C/3D site. A $I_{74} \rightarrow T$ conversion in the poliovirus 3C proteinase also exhibits decreased 3C/3D cleavage activity (KEAN et al. 1988). In addition, a slight decrease in the cleavage at the 2B/2C junction was observed. The effects of these mutations on the virus infectious cycle will be discussed further below.

It has become evident that the system in which a particular mutation is evaluated (expression of cDNA in bacteria; in vitro translation of RNA transcribed from cDNAs; or in cell culture) as well as how the enzyme is supplied (in trans from subgenomic clones and infected cell extracts or in cis as part of a polyprotein) can greatly influence the eventual interpretation of the effect on processing. Different mutations at the same site also cause different phenotypes. Future studies of processing should include as many substrates as possible and evaluate both the trans and cis activities of the proteinases.

4.1.2 Lesions of 3CD

The demonstration of 3CD as the minimal P3 protein capable of complete P1 processing in poliovirus (JORE et al. 1988; YPMA-WONG et al. 1988b) and the activity of EMCV precursors toward 3C substrates (PARKS et al. 1989) has indicated a unique mechanism utilized by picornaviruses in the regulation of gene expression. Proteolytic activity may be regulated by both the structure of the substrates and the form in which the processing activity is provided. The insertion of four amino acids at aa_{131} of the 3D protein disrupted P1 processing in an in vitro translation system (YPMA-WONG et al. 1988b). Mutational analysis of 3D was expanded by BURNS et al. (1989) who analyzed the effect of several 3D mutations on polymerase activity and 3CD-mediated proteolysis of P1. All of the mutations affected polymerase activity. Cleavage of the 3C/3D bond was only mildly affected in all mutants described, as assayed by bacterial expression of subgenomic clones encoding only the 3CD protein. Again, as in the EMCV and PV1 insertion mutations discussed above, the data suggest that only a limited region of the protein serves in recognition as a substrate. When the *trans* cleavage activity of the 3CD mutants was analyzed by addition of bacterial extracts containing mutant 3CD (as well as 3C and 3D) to in vitro translation reactions containing P1 protein, it was seen that insertion of an amino acid at aa_{149} and at aa_{290} did not disrupt processing of P1 whereas insertion of single amino acid at position 241, 257, 236, and insertion of four amino acids at position 147 in the 3D portion of 3CD prevented complete processing. No functional correlation between the location of the mutations and the resultant processing phenotype could be found. Although several more mutations of 3D are known, the effects of the mutations on P1 processing have not been assessed (AGUT et al. 1989; BELLOCQ et al. 1987; BERNSTEIN et al. 1986; KEAN et al. 1988; LEE and WIMMER 1988; PLOTCH et al. 1989; TOYODA et al. 1987). Replacement of the CVB3 3D coding sequences with those of PV1 severely limits the processing activity of coxsackievirus B3 3C on P1 in in vitro translation assays (M. A. LAWSON et al., submitted for publication). The processing of EMCV P1 has recently been shown to be carried out by larger precursors containing 3C sequences as well as by 3C itself (PARKS et al. 1989). The differential cleavage activity of 3C and 3CD is not seen in EMCV processing. It should also be noted that truncation of the FMDV 3D protein is not detrimental to P1 processing in a rabbit reticulocyte lysate translation reaction (CLARKE and SANGER 1988).

4.1.3 Lesions of 2A

In demonstrating the proteolytic activity of 2A in PV3, TOYODA et al. (1986) used a combination of truncation and mutation of the 2A coding region to disrupt processing at the P1/2A junction. The mutations used were a four amino acid insertion proximal to the amino terminus of 2A and a nine amino acid deletion proximal to the carboxy terminus. Expression of the above mutant cDNAs in an in vitro translation system revealed that the four amino acid insertion was capable of P1/2A cleavage activity. Whereas the nine amino acid deletion mutant was incapable of P1/2A processing, it was competent as a substrate for exogenous wild type 2A activity (NICKLIN et al. 1987). A six amino acid insertion into 2A position 67 (YPMA-WONG et al.1988a) exhibits a retarded rate of cleavage of the P1/2A junction (M. A. LAWSON and B. L. SEMLER, unpublished observations). Whether the p220 cleavage is directly dependent on the proteolytic activity of the enterovirus and rhinovirus 2A proteinases remains to be elucidated. It should also be noted that ROOS et al. (1989a, b) have reported the isolation of a four amino acid insertion of the TMEV 2A coding region. Such a lesion appeared to have little or no effect on in vitro protein processing and did not interfere with the production of virus when synthetic RNAs derived from full-length cDNAs bearing the mutation were transfected into cultured mouse cells. Further mutational analyses will be required to define the precise function of 2A in TMEV.

4.1.4 Lesions of the FMDV L Proteinase

Both the cardiovirus and aphthovirus genomes encode an additional protein at the 5' end of the polyprotein coding region which has been termed the leader or L protein. The L protein of the aphthovirus FMDV has proteolytic activity and is responsible for the L/P1 cleavage event (STREBEL and BECK 1986). The role of the cardiovirus EMCV and TMEV L proteins is unclear, as the cleavage of the EMCV L/P1 junction has been shown to be catalyzed by the 3C proteinase (PARKS et al. 1986). The L protein of FMDV C_1O strain was mutagenized by STREBEL and BECK (1986) using the bisulfite method (KALDERON et al. 1982). A double bearing $D_{35} \rightarrow N$ and $R_{61} \rightarrow K$ changes and three mutants bearing a $G_{36} \rightarrow D$, a $T_{55} \rightarrow I$ or a $P_{66} \rightarrow S$ change were recovered. Only the $T_{55} \rightarrow I$ mutant was incapable of L/P1 cleavage upon bacterial expression of a cDNA bearing the lesion or in in vitro translation reactions programmed with RNA encoding the mutation.

4.2 Lesions That Affect Substrate Recognition

Distinguishing between mutants in which the enzymatic activity of a proteinase is affected rather than recognition of the substrate can be difficult when the enzymes themselves are also substrates for their own cleavage activity. In resolving this problem, the utility of testing proteinase mutants as substrates for wild-type proteinase activity has been helpful, but the problem of assigning the effects of a mutation on catalysis or on substrate recognition requires careful biochemical examination which is not always possible to carry out due to the lack of purified enzymes and substrates.

Viral proteins that are only cleavage substrates and do not bear any proteolytic activity themselves do not present this obstacle. The P1 precursor has been widely exploited to analyze the differential activities of viral proteinases on various cleavage sites. The activity of 3C has been most thoroughly studied using P1 as a substrate. That the structural integrity of the P1 precursor must be maintained to serve as a substrate for 3C or 3CD activity was first alluded to in an in vitro translation study in which a portion of the 1D and 2A proteins were deleted, preventing processing of the P1 precursor into 1AB, 1C, and 1D (YPMA-WONG and SEMLER 1987a). A truncation of the P1 precursor also led to a loss of processing by 3C (NICKLIN et al. 1987). A more detailed truncation study of P1 (YPMA-WONG and SEMLER 1987b) showed that nearly the entire P1 protein was necessary for processing to occur, indicating the involvement of highly ordered structural determinants in substrate recognition. A report by CLARKE and SANGAR (1988) described the deletion of the L, 1A, and a portion of the 1B proteins of FMDV in a subgenomic cDNA clone from which RNA was then synthesized and used to program translation in a rabbit reticulocyte lysate. Their results showed that such a deletion prevented utilization of P1 as a substrate by 3C activity. The presence of complete 3D sequences was not found to be required for normal processing of FMDV P1.

The requirements for highly ordered structures in 3CD-mediated processing of the poliovirus type 1 P1 precursor have been examined in depth. The availability of the crystal structure of the mature poliovirion (HOGLE et al. 1985) has allowed a better assessment of how particular mutations may affect the structure of the substrate P1 molecule. Analysis of several insertion and deletion mutants in the P1 protein by in vitro translation and cleavage assays has shown a clear correlation between the site of the lesion in the context of the mature virion structure and the efficiency of processing of the P1 protein (YPMA-WONG et al. 1988a). Mutations in the highly ordered β-barrel core of the capsid proteins were found to be more detrimental to processing than those in intervening loop regions. Although the actual structure of P1 is not known, the known structure of the mature virion does provide valuable insight into the mechanisms by which the P1 precursor is processed. Of particular interest is a mutant P1 protein in which an additional Q-G cleavage site was inserted adjacent to the authentic Q-G site at the 1C/1D junction. The new cleavage site is preferentially cleaved by 3CD activity, although cleavage of the original site can also be detected. Cloning of the mutant P1 protein into a full-length PV1 cDNA and transfection of the mutant cDNA onto HeLa cells yielded viable virus that exhibited secondary morphogenetic defects (BLAIR et al. 1990). The implications for substrate recognition will be discussed further below.

Turning to finer detail in substrate recognition, the nature of the dipeptide sequence required for cleavage of the 1C/1D junction of EMCV has been investigated by PARKS and PALMENBERG (1987). Site directed mutagenesis of the Q-G cleavage site was used to convert the Q-G dipeptide sequence to Q-A, Q-E, Q-V, K-G, K-A, K-V, and K-E. It was found that only the Q-A (P_1-P'_1) sequence still allowed processing at the 1C/1D junction. A subsequent study involved modification of the 3B/3C and 3C/3D junctions of EMCV (PARKS et al. 1989). The Q-G sequences were modified to R-G, L-G, and to Q-I, Q-C, Q-T, or Q-Y for the 3B/3C and 3C/3D junctions respectively. It was found that only the Q-C allowed processing by 3C activity. These mutations did not appear to affect P1 processing activity. When one includes the E-S and Q-S dipeptide pairs that occur naturally in the EMCV genome and are cleaved by 3C activity, the limited variability of sequences becomes significant and suggests that certain character-istics are required of dipeptide sequences to function in catalysis.

A study of the cleavage activity of purified PV2 3C on synthetic peptides representing the Q-G cleavage sites within the polyprotein has revealed a significant variability in the rate of cleavage of the Q-G pairs. The data from this study demonstrate a clear role of the primary sequence surrounding the cleaved Q-G bond in substrate utilization. The rates of cleavage of peptides do not mimic the rates of cleavage seen in infected cells in all cases, providing evidence for other factors such as higher order structures in substrate recognition (PALLAI et al. 1989). Such analyses should continue to provide valuable information on the mechanism of catalysis by the picornavirus 3C proteinase and on how the regulation of processing by 3C is controlled.

Other studies of substrate recognition have involved P3 proteins. A duplication of S_{182} in the PV1 3C has the effect of moving the P_4 aliphatic T residue to P_5 and shifts the P3 Q into the P_4 position. Expression of a subgenomic cDNA clone containing the mutation in *E. coli* showed that the duplication prevented processing at the 3C/3D junction. Transfection of this clone onto COS-1 or HeLa cells did not yield virus (SEMLER et al. 1987). The above observation was later confirmed following in vitro translation of full length in vitro transcribed RNA of a poliovirus type 1 cDNA encoding the lesion (YPMA-WONG et al. 1988b). The processing at the 3C/3D site was eliminated (as predicted by the bacterial expression experiments), and the 3CD-mediated processing of P1 was nearly identical to that of wild type.

A more recent application of cassette mutagenesis technology has been used in a functional analysis of the viral genome-linked protein 3B of PV1 (KUHN et al. 1988a, b). Several mutant cDNAs were made, some of which produced viable virus upon transfection of in vitro synthesized RNA onto HeLa cell monolayers. A general phenotype among the viable viruses was an alteration of in vitro processing of 3B-containing proteins. Among the nonviable mutant cDNAs, the production of 3CD from in vitro transcribed mRNAs was generally affected, leading to the secondary phenotype of inefficient or undetectable P1 processing. It should be noted that not all nonviable 3B mutations made in these studies were impaired in processing (KUHN et al. 1988b).

The Y-G cleavage site in the poliovirus type 1 3D protein which is cleaved by 2A to yield 3C' and 3D' has also been studied (LEE and WIMMER 1988). The P_1 Y_{148} residue was altered by site directed mutagenesis to F and the $P_2 T_{147}$ was changed to A. Both mutations yielded virus with wild type growth properties upon DNA transfection onto COS-1 monolayers. Interestingly, the $T_{147} \rightarrow A$ conversion eliminated processing at the 3C'/3D' site whereas the $Y_{148} \rightarrow F$

conversion continued to be processed. The implication of this finding is that neither the 3C' nor 3D' proteins are required for efficient growth in cell culture.

4.3 Lesions That Affect Other Viral Functions

Because of the intimate interaction among picornaviral proteins, it is probable that mutations in one protein can exhibit a global effect on the virus life cycle. An example of this can be found in studies of a leucine insertion between G_{102} and F_{103} of the PV1 2A protein which resulted in an infectious virus upon transfection of mutant cDNA onto CV1 cells (BERNSTEIN et al. 1985). The effect of the mutation on the efficiency of P1/2A cleavage was not investigated. The mutation did result in the loss of p220 cleavage activity and caused a variety of protein and RNA synthesis defects presumably as the result of inefficient host cell shut off. The differential effect of mutations in 2A on p220 cleavage has been more recently reported by KRÄUSSLICH et al. (1987) who demonstrated that a four amino acid insertion mutation and a nine amino acid deletion mutation in the 2A protein abrogated p220 cleavage activity. As mentioned above, the four amino acid insertion mutant was still capable of processing the P1/2A junction, albeit less efficiently.

Viruses that have been recovered from cassette mutagenesis of 3C have exhibited small plaque phenotypes and RNA synthesis defects by possibly altering the function of protein 3CD in replication (DEWALT and SEMLER 1987, 1989). The mutant virus Se1-3C-02 bearing a $V_{54} \rightarrow A$ conversion was temperature-sensitive for RNA synthesis at 39 °C even though its altered protein processing phenotype was not exacerbated at the nonpermissive temperature. DEWALT and SEMLER (1989) also showed that the Se1-3C-02 mutant virus could be rescued by coinfection with a chimeric poliovirus (JOHNSON and SEMLER 1988) bearing the 5' noncoding region of CVB3. The $I_{74} \rightarrow T$ conversion in 3C mentioned above also exhibits a deficiency in RNA synthesis, but whether the deficiency was in total synthesis or in delayed kinetics of synthesis was not clear. The mutant virus also exhibited a temperature sensitive defect in total yield of infectious virions at 39 °C (KEAN et al. 1988). A chimeric poliovirus bearing the 3B sequence of echovirus 9 (W1-VPg-29) and a mutant poliovirus bearing two 3B lesions (W1-VPg-34) were found to have delayed growth kinetics as measured by the one step growth experiment. The proteinases produced by mutant virus W1-VPg-34 process P1 less efficiently in in vitro translation assays than those produced by the wild-type virus (KUHN et al. 1988b).

The above examples of mutational analysis of processing in picornaviruses show a wide variety of phenotypes, giving evidence for host cell-virus interactions that have yet to be characterized. Further studies of the existing mutants and creation of new mutations in the virus polyprotein will be invaluable to our discovery and understanding of the regulation of gene expression of picornaviruses and the interactions of the viruses with host cells.

5 Biochemical Implications of the Observed Picornavirus Protein Processing Activities

5.1 Primary Site Recognition

The overall conservation of cleavage sites recognized by the proteinases of individual picornaviruses suggests that the sites themselves are primary determinants of proteolytic processing. For example, the 3C proteinase of poliovirus cleaves exclusively at Q-G pairs within its wild-type polyprotein sequence. Such strict specificity of sites cleaved appears to be more the exception than the rule when the known and proposed cleavage sites recognized by other picornavirus 3C proteinases are examined. The 3C proteinase of EMCV cleaves at specific Q-G, Q-S, and possibly E-S sites while the 3C activity of FMDV appears to cleave at E-G, E-S, V-G, Q-T, Q-L, and Q-I amino acid pairs. Indeed, as mentioned above PARKS and PALMENBERG (1987) demonstrated that some flexibility exists in which amino acid pairs can be recognized and cleaved by 3C. Using site-directed mutagenesis of cDNA clones expressing the EMCV P1 precursor, they showed that alanine could substitute for glycine in the Q-G pair between VP3 and VP1. These investigators also reported that there were a number of amino acid replacement for both residues at the VP3-VP1 cleavage site that could not be cleaved in their in vitro assay. Thus it appears that the 3C proteinases of picornaviruses are able to recognize and cleave a limited set of amino acid pairs distinct from those actually found at wild-type cleavage sites.

An additional example of site selection flexibility is the 2A proteinase of poliovirus. LEE and WIMMER (1988) used oligonucleotide-directed mutagenesis to alter the Y-G cleavage site normally cleaved to produce 3C′ and 3D′. Substitution of F for Y at the P_1 site did not affect cleavage, demonstrating that the poliovirus 2A proteinase can cleave amino acid pairs other than those which are normally cleaved during wild-type virus infections.

The site specificity of picornavirus proteinases appears to extend beyond the P_1-P_1' amino acid residues that are actually cleaved by the individual enzymes. As originally noted by NICKLIN et al. (1986), there is a consensus of an alanine residue in the P_4 position of sites cleaved by the poliovirus 3C proteinase activity. Interestingly, the P_4 residue at the polio 3C/3D junction is a threonine residue and cleavage at this site is infrequent (compared to, for example, the 3B/3C junction) both in vivo and in vitro (DEWALT and SEMLER 1987; YPMA-WONG et al. 1988b). In a recent report employing synthetic peptides as substrates for purified poliovirus 3C, PALLAI et al. (1989) changed the P_4 threonine residue to an alanine within a 16 amino acid residue peptide that mimics the 3C/3D cleavage site. Such an alteration resulted in a dramatic increase in cleavage efficiency that was nearly equal to the efficiency of cleavage of a peptide that mimics the 3B/3C cleavage site (that contains a P_4 alanine as part of its wild-type sequence). In addition, YPMA-WONG et al. (1988a) have generated amino acid insertion mutations in in vitro synthesized P1 polypeptides of poliovirus and have shown that an inserted Q-G pair with a P_4 alanine residue is cleaved almost exclusively over the wild-type Q-G site (six amino acids downstream) whose P_4 alanine was changed to a glycine residue as a result of the amino acid insertion. It was suggested that, while not a processing determinant in itself, the P_4 residue at a Q-G cleavage site may influence the stability of the enzyme-substrate complex and increase the efficiency of cleavage.

An extended amino acid recognition sequence for the poliovirus 2A proteinase was recently demonstrated by the molecular genetic experiments of LEE and WIMMER (1988), as mentioned above. It was observed that poliovirus strains that synthesize detectable quantities of 3C' and 3D' in infected cells encode a threonine residue in the P_2 position of their Y-G cleavage sites, while those that do not produce 3C' and 3D' contain an alanine in the P_2 position. In addition, all of the VP1-2A junction sequences of different polioviruses contain a threonine residue in the P_2 position of their Y-G cleavage sites. When the amino acid residue in the P_2 position of the PV1 (Mahoney) 3C'/3D' cleavage site was mutated to an alanine residue, no cleavage at the Y-G pair was detected and, hence, no 3C' and 3D' polypeptides were produced. It will be of interest to determine whether mutation of the P_2 amino acid of the VP1/2A junction sequence to an alanine results in an alteration of cleavage efficiency, since cleavage at this Y-G pair is normally very efficient and is thought to occur while polyprotein synthesis is still taking place on cellular ribosomes. It will also be important to determine the nature, if any, of the extended recognition sequence that may exist at the 2A/2B junction of the EMCV, FMDV, and TMEV polyproteins. Knowledge of such a sequence may aid in the identification of the enzyme(s) responsible for cleavage at the above site for these different picornaviruses.

5.2 Conformational Contributions of Precursor Polypeptides

As detailed in Sect. 3 above, the use of amino acid analogue incorporation and temperature-shift experiments during a picornavirus infection established that the conformation of substrate precursor polypeptides is an important determinant in protein processing. ARNOLD et al. (1987) used the crystallographic data obtained for the HRV14, PV1, and Mengo virus capsids to analyze the interaction of the 3C proteinase with the P1 (capsid) precursor polypeptide. These studies suggested that, in addition to surface accessibility, authentic cleavage sites need to be presented in a structurally flexible context in order to be recognized by the 3C (or 3CD) proteinase. Such flexibility may be required for correct positioning of the scissile bond of the substrate within the active site of the enzyme. Indeed, an examination of the structural context of amino acid pairs within the P1 precursor that are not used as 3C cleavage sites (but are among the dipeptides normally recognized by the 3C enzyme for a specific picornavirus) reveals that such sites are usually found within highly ordered structural domains (ARNOLD et al. 1987).

Experimental evidence for the role of structural domains in the cleavage of picornavirus capsid precursors by the 3C proteinase activity comes from engineering truncations or site-directed alterations into the P1 precursor polypeptide and testing such mutated proteins as substrates for in vitro processing activities (YPMA-WONG and SEMLER 1987a, b; NICKLIN et al. 1987; YPMA-WONG et al. 1988a; CLARKE and SANGAR 1988). Collectively, the results from these studies suggest that the viral 3C (and 3CD) proteinases recognize their cognate amino acid pairs within the context of a properly folded precursor polypeptide. It is of interest that the above conclusions are based upon the use of the P1 precursor (whose highly ordered overall structure can be extrapolated from the known structures of picornavirus particles) as a substrate for 3C activity. Alterations (primarily by in-phase amino acid insertions) in the precursors to nonstructural proteins have proved less detrimental in their effects on these precursors to serve as authentic 3C substrates (PARKS et al. 1986, 1989; YPMA-WONG and SEMLER 1987a, b; ROOS et al. 1988a). Additional evidence for different 3C-substrate interactions for structural vs nonstructural precursors comes from recent experiments by DEWALT et al. (1989), in which chimeric polyproteins were generated containing 3C sequences from HRV14 or CVB3 in a PV1 background. The results of in vitro processing experiments demonstrated that proteolytic processing of poliovirus nonstructural precursor polypeptides (i.e., P2 and P3) could be carried out by the heterologous proteinases, even at some amino acid pairs not normally cleaved by that proteinase. In contrast, the heterologous 3C proteinases expressed from the chimeric genomes were incapable of recognizing the poliovirus specific processing sites within the structural (P1) precursor. The processing of the P1 capsid precursor thus appears to be more virus-specific than that of the P2 and P3 region precursors. Overall, the cross-species processing experiments demonstrate that there are common conformational determinants, in addition to specific amino acid pairs, that are necessary for 3C-specific processing.

5.3 Proteolytic Activity of Precursor Polypeptides Containing the 3C Amino Acid Sequence

The generation of the 3C polypeptide itself must occur by an intramolecular, self-cleavage reaction (PALMENBERG and RUECKERT 1982; HANECAK et al. 1984) or by the activity of one P2-P3 precursor on another because the protein is flanked by its own cleavage recognition sites. The production of 3C by these cleavage activities requires that processing activity be extant in precursor polypeptides. The generation of mutated and defined P3 region polypeptides by in vitro translation of transcripts derived from genetically engineered poliovirus and EMCV cDNAs has provided direct proof that such precursor proteins have 3C-specific proteinase activity (YPMA-WONG et al. 1988b; JORE et al. 1988; PARKS et al. 1989). The above observations have at least two important consequences during a picornavirus infection. First, although perhaps only limited to poliovirus and coxsackievirus, the requirement for 3D sequences (in the form of protein 3CD) for efficient P1 processing allows the virus to generate different processing efficiencies for different viral substrates based upon the same core proteinase activity (i.e., mature 3C). The additional 3D sequences may catalyze the formation of a stable P1-enzyme complex in which the Q-G pairs are correctly positioned for interaction with the proteinase (catalytic) domain of 3CD. In support of such a proposal, NICKLIN et al. (1988) have shown that highly purified 3C (expressed from genetically engineered *E. coli*) is very inefficient at in vitro cleavage of the VP0-VP3 bond (in the poliovirus P1 precursor) compared to cleavage of the VP3-VP1 Q-G pair and that the overall P1 processing is much less efficient than that catalyzed by 3CD provided by an infected cell extract.

A second consequence of generating picornavirus precursor polypeptides with 3C-specific proteolytic activity is that, during early times after initiating an infection, picornaviruses are not dependent on the cleavage activity of a single, mature polypeptide (3C) whose production in significant quantities is usually only seen at relatively late times after infection. This latter observation has been confirmed for EMCV by a detailed kinetic analysis of in vitro translation of the viral genome and subsequent protein processing events (JACKSON 1986). The in vitro studies provided convincing evidence that 3ABC, rather than 3C itself, carries out the early stages of processing the EMCV capsid precursor (P1), perhaps because the cleavage of the carboxy-terminal Q-G pair of EMCV 3C is much more efficient than cleavage at its amino-terminal Q-G site. For poliovirus, it was first noted that the amino-terminal cleavage of 3C was much more rapid than the carboxy-terminal Q-G cleavage in bacteria expressing a subgenomic cDNA of poliovirus containing 3C sequences (HANECAK et al. 1984). As confirmed by the subsequent in vitro studies mentioned above, the result of such differential cleavage rates of the Q-G sites flanking poliovirus 3C is that the virus uses polypeptide 3CD as a major proteinase that is capable of processing structural and nonstructural precursor polypeptides. Perhaps the picornaviruses have evolved different cleavage efficiencies for the sites flanking their respective 3C proteinases (via conformational or surrounding amino acid determinants) as a means of insuring faithful processing at both early and late times after infection and as a means of generating proteinase-containing polypeptides with differential catalytic activities in *cis* vs *trans*. This latter topic is discussed in next section.

5.4 The Importance of *Cis* vs *Trans* Cleavage

A final biochemical consideration that results from the cleavage specificities and precursor activities of picornavirus proteinases is the contribution of intra-molecular versus intermolecular processing events. To date, three picornavirus proteinases have been identified as having intramolecular (i.e., cleavage in *cis*) protein processing activities: (a) the L protein of FMDV (STREBEL and BECK 1986); (b) the 2A protein of poliovirus (TOYODA et al. 1986); and (c) the 3C protein of EMCV (PALMENBERG and RUECKERT 1982) and PV (HANECAK et al. 1984). As mentioned in Sect. 2.1 above, during the early phases of a picornavirus infection the relative concentrations of polyproteins per cell may be low enough that *trans* cleavage events (i.e., bimolecular reactions) are rare. In this case, the virus would depend on the intramolecular cleavage of precursor polypeptides containing L, 2A, or 3C to liberate diffusable proteinases or to cleave the P2-P3 fusion precursor in *cis* at sites that would produce polypeptides required for RNA replication (e.g., 3D, 2C, or 3CD). For example, certain sites within the P2-P3 region of the polyprotein may be particularly good substrates for intramolecular cleavage in order to generate polypeptides absolutely essential for early replication events. Other sites may be more accessible for *trans* cleavage by a free, diffusable polypeptide containing 3C proteinase activity. Cleavage of specific sites in *trans* could produce proteins that are not essential early in the infectious cycle. The 3C (or 3CD)-mediated cleavage of the capsid precursor (P1) appears to be an example of this latter class of cleavage sites. The use of site-directed mutagenesis to produce picornaviruses with conditional processing lesions will directly confirm whether a strict correlation exists between temporal expression of distinct viral gene functions and the *cis* versus *trans* nature of the cleavage events that produce the proteins responsible for execution of such functions.

6 Summary and Future Directions

Our understanding of the regulation of picornavirus gene expression by proteolytic processing has increased considerably since evidence for processing of precursor polypeptides was first reported more than 20 years ago. Since that time, we have learned about the precursor-to-product relationships of many viral-specific polypeptides. We have used critical information stored in the RNA sequences known for nearly all picornaviruses to determine the sites of proteolytic cleavage and to create a precise genetic map of the different viral genomes. The discovery of at least three different proteinase activities has underscored both the similarities and differences in the processing strategies used by different picornaviruses. In addition, genetic manipulation of cloned viral cDNAs has produced some of the first glimpses of the nature of the enzyme-substrate interactions that occur between picornavirus proteinases and the polypeptide precursors that they ultimately cleave. However, future experiments using purified enzymes and substrates will refine and expand the rather rudimentary knowledge that we have concerning both the catalytic domains of the viral proteinases and the precise residues that are contacted in substrate polypeptides. Knowledge of the site-recognition domains of picornaviral proteinases will be a key step in uncovering the mechanisms of the dynamic and selective cleavage events involved in the complete processing of viral polyproteins. These studies will be greatly aided by the ongoing attempts to solve the three-dimensional structure of the picornavirus proteinases and to correlate such structures with the known and predicted processing phenotypes for the different enzymes being studied. The use of genetically defined viral genomes in both in vitro and in vivo studies will provide the ultimate tests for the precise regulatory roles of the viral proteinases during the course of an infection.

References

Abraham G, Cooper PD (1975) Poliovirus polypeptides examined in more detail. J Gen Virol 29: 199–213

Acharya R, Fry E, Stuart D, Fox G, Rowlands D, Brown F (1989) The three-dimensional structure of foot-and-mouth disease virus at 2.9 Å resolution. Nature 327: 709–716

Adler CJ, Elzinga M, Wimmer E (1983) The genome-linked protein of picornaviruses. VII. Complete amino acid sequence of poliovirus VPg and carboxy-terminal analysis of its precursor, P3–9. J Gen Virol 64: 349–355

Agut H, Kean KM, Fichot O, Morasco J, Flanegan JB, Girard M (1989) A point mutation in the poliovirus polymerase gene determines a complementable temperature-sensitive defect of RNA replication. Virology 168: 302–311

Ambros V, Baltimore D (1978) Protein is linked to the 5' end of poliovirus RNA by a phosphodiester linkage to tyrosine. J Biol Chem 253: 5263–5266

Ambros V, Pettersson RF, Baltimore D (1978) An enzymatic activity in uninfected cells that cleaves the linkage between poliovirion RNA and the 5' terminal protein. Cell 15: 1439–1446

Andrews NC, Baltimore D (1986) Purification of a terminal uridylyltransferase that acts as host factor in the in vitro poliovirus replicase reaction. Proc Natl Acad Sci USA 83: 221–225

Argos P, Kramer G, Nicklin MJH, Wimmer E (1984) Similarity in gene organization and homology between proteins of animal picornaviruses and a plant comovirus suggest common ancestry of these virus families. Nucleic Acids Res 12: 7251–7267

Armstrong JA, Edmonds M, Nakazato H Phillips BA, Vaughan MH (1972) Polyadenylic acid sequences in the virion RNA of poliovirus and eastern equine encephalitis virus. Science 176: 526–528

Arnold E, Luo M, Vriend G, Rossmann MG, Palmenberg AC, Parks GD, Nicklin MJH, Wimmer E (1987) Implications of the picornavirus capsid structure for polyprotein processing. Proc Natl Acad Sci USA 84: 21–25

Bachrach, HL, Swaney JB, Vande Woude GF (1973) Isolation of the structural polypeptides of foot-and-mouth disease virus and analysis of their C-terminal sequences. Virology 52: 520–528

Baltimore D (1971) Polio is not dead. In: Pollard M (ed) Perspectives in virology, vol 1. Academic, New York, pp 1–14

Baltimore D, Franklin RN, Eggers HJ, Tamm I (1963) Poliovirus induced RNA polymerase and the effects of virus-specific inhibitors on its production. Proc Natl Acad Sci USA 49: 843–849

Baltimore D, Jacobson MF, Asso J, Huang AS (1969) The formation of poliovirus proteins. Cold Spring Harb Symp Quant Biol 34: 741–746

Bazan JF, Fletterick RJ (1988) Viral cysteine proteases are homologous to the trypsin-like family of serine proteases: structural and functional implications. Proc Natl Acad Sci USA 85: 7872–7876

Bellocq C, Kean KM, Fichot O, Girard M, Agut H (1987) Multiple mutations involved in the phenotype of a temperature-sensitive small-plaque mutant of poliovirus. Virology 157: 75–82

Bernstein HD, Sonenberg N, Baltimore D (1985) Poliovirus mutant that does not selectively inhibit host cell protein synthesis. Mol Cell Biol 5: 2913–2923

Bernstein HD, Sarnow P, Baltimore D (1986) Genetic complementation among poliovirus mutants derived from an infectious cDNA clone. J Virol 60: 1040–1049

Blair WS, Hwang SS, Ypma-Wong MF, Semler BL (1990) A mutant poliovirus containing a novel proteolytic cleavage site in VP3 is altered in viral maturation. J Virol 64: 1784–1793

Bonneau A, Sonenberg N (1987) Proteolysis of the p220 component of the cap-binding complex is not sufficient for complete inhibition of host cell protein synthesis after poliovirus infection. J Virol 61: 986–991

Boothroyd JC, Harris TJR, Rowlands DJ, Lowe PA (1982) The nucleotide sequence of cDNA coding for the structural proteins of foot-and-mouth disease virus. Gene 17: 153–161

Burns CC, Lawson MA, Semler BL, Ehrenfeld E (1989) Effects of mutations in poliovirus 3D^{pol} on RNA polymerase activity and on polyprotein cleavage. J Virol (in press)

Burrell CJ, Cooper PD (1973) N-Terminal aspartate, glycine, and serine in poliovirus capsid proteins, J Gen Virol 21: 443–451

Burroughs JN, Sangar DV, Clarke BE, Rowlands DJ, Billiau A, Collen D (1984) Multiple proteases in foot-and-mouth disease virus replication. J Virol 50: 878–883

Butterworth BE (1973) A comparison of the virus-specific polypeptides of encephalomyocarditis virus, human rhinovirus-1A, and poliovirus. Virology 56: 439–453

Butterwoth BE, Korant B (1974) Characterizaion of the large picornaviral polypeptides produced in the presence of zinc ion. J Virol 14: 282–291

Butterworth BE, Rueckert RR (1972a) Gene order of encephalomyocarditis virus as determined by studies with pactamycin. J Virol 9: 823–828

Butterworth BE, Rueckert RR (1972b) Kinetics of synthesis and cleavage of encephalomyocarditis virus-specific proteins. Virology 50: 535–549

Butterworth BE, Hall L, Stoltzfus CM, Rueckert RR (1971) Virus specific proteins synthesized in encephalomyocarditis virus-infected HeLa cells. Proc Natl Acad Sci USA 68: 3083–3087

Caliguiri LA, Tamm I (1970) The role of cytoplasmic membranes in poliovirus biosynthesis. Virology 42: 100–111

Callahan PL, Mizutani S, Colonno RJ (1985) Molecular cloning and complete sequence determination of the RNA genome of human rhinovirus type 14. Proc Natl Acad Sci USA 82: 732–736

Carroll AR, Rowlands DJ, Clarke BE (1984) The complete nucleotide sequence of the RNA coding for the primary translation product of foot and mouth disease virus. Nucleic Acids Res 12: 2461–2472

Celma ML, Ehrenfeld E (1975) Translation of poliovirus RNA in vitro: detection of two different initiation sites. J Mol Biol 98: 761–780

Clarke BE, Sangar DV (1988) Processing and assembly of foot-and-mouth disease virus proteins using subgenomic RNA. J Gen Virol 69: 2313–2325

Clarke BE, Sangar DV, Burroughs JN, Newton SE, Carroll AR, Rowlands DJ (1985) Two initiation sites for foot-and-mouth disease virus polyprotein in vivo. J Gen Virol 66: 2615–2626

Craik CS, Rozniak S, Sprang S, Fletterick R, Rutter W (1987) Redesigning trypsin via genetic engineering. J Cell Biochem 33: 199–211

Dasgupta A, Zabel P, Baltimore D (1980) Dependence of the acitivity of the poliovirus replicase on a host cell protein. Cell 19: 423–429

Devaney MA, Vakharia VN, Lloyd RE, Ehrenfeld E, Grubman MJ (1988) Leader protein of foot-and-mouth disease virus is required for cleavage of the p220 component of the cap-binding complex. J. Virol. 62: 4407–4409

Dewalt PG, Semler BL (1987) Site directed mutagenesis of proteinase 3C results in a poliovirus deficient in synthesis of viral RNA polymerase. J Virol 61: 2162–2170

Dewalt PG, Semler BL (1989) Molecular biology and genetics of poliovirus protein processing. In: Semler BL, Ehrenfeld E (eds) Molecular aspects of picornavirus infection and detection. American Society for Microbiology, Washington DC, pp 83–93

Dewalt PG, Lawson MA, Colonno RJ, Semler BL (1989) Chimeric picornavirus polyproteins demonstrate a common 3C proteinase substrate spcificity. J Virol 63: 3444–3452

Dewalt PG, Blair WS, Semler BL (1990) A genetic locus in mutant poliovirus genomes involved in overproduction of RNA polymerase and 3C proteinase. Virology 174: 504–514

Doel TR, Sangar DV, Rowlands DJ, Brown F (1978) A re-appraisal of the biochemical map of foot-and-mouth disease virus RNA. J Gen Virol 41: 395–404

Domier LL, Shaw JG, Rhoads RE (1987) Poxyviral proteins share amino acid sequence homology with picorna-, como-, and caulimoviral proteins. Virology 158: 20–27

Dorner AJ, Semler BL, Jackson RJ, Hanecak R, Duprey E, Wimmer E (1984) In vitro translation of poliovirus RNA: utilization of internal initiation sites in reticulocyte lysate. J Virol 50: 507–514

Emini EA, Elzinga M, Wimmer E (1982) Carboxy-terminal analysis of poliovirus proteins: termination of poliovirus RNA translation and location of unique poliovirus polyprotein cleavage sites. J Virol 42: 194–199

Etchison D, Fout S (1985) Human rhinovirus 14 infection of HeLa cells results in the proteolytic cleavage of the p220 cap-binding complex subunit and inactivates globin mRNA translation in vitro. J Virol. 54: 634–638

Etchison D, Milburn SC, Edery I, Sonenberg N, Hershey JWB (1982) Inhibition of HeLa cell protein synthesis following poliovirus infection correlates with the proteolysis of a 220,000-dalton polypeptide associated with eucaryotic initiation factor 3 and a cap binding protein complex. J Biol Chem 257: 14806–14810

Etchison D, Hansen J, Ehrenfeld E, Edery I, Sonenberg N, Milburn S, Hershey JWB (1984) Demonstration in vitro that eukaryotic initiation factor 3 is active but that a cap binding protein complex is inactive in poliovirus-infected HeLa cells. J Virol 51: 832–837

Flanegan JB, Pettersson RF, Ambros V, Hewlett MJ, Baltimore D (1977) Covalent linkage of a protein to a defined nucleotide sequence at the 5′-terminus of virion and replicate intermediate RNAs of poliovirus. Proc Natl Acad Sci USA 74: 961–965

Forss S, Strebel K, Beck E, Schaller H (1984) Nucleotide sequence and genome organization of foot-and-mouth disease virus. Nucleic Acids Res 12: 6587–6601

Franssen H, Leunissen J, Goldbach R, Lomonossoff G, Zimmern D (1984) Homologous sequences in non-structural proteins from cowpea mosaic virus and picornaviruses. EMBO J 3: 855–861

Garfinkle BD, Tershak DR (1971) Effect of temperature on the cleavage of polypeptides during growth of LSc poliovirus. J Mol Biol 59: 537–541

Girard M, Baltimore D, Darnell JE (1967) The poliovirus replication complex; site for synthesis of poliovirus RNA. J Mol Biol 24: 59–74

Gorbalenya AE, Svitkin YV (1983) Encephalomyocarditis virus protease: purification and role of the SH groups in processing of the precursor of structural proteins. Biochemistry (USSR) 48: 385–395

Gorbalenya AE, Svitkin YV, Kazachkov YA, Agol VI (1979) Encephalomyocarditis virus-specific polypeptide p22 is involved in the processing of the viral precursor polypeptides. FEBS Lett 108: 1–5

Gorbalenya AE, Svitkin YV, Agol VI (1981) Proteolytic activity of the nonstructural polypeptide p22 of encephalomyocarditis virus. Biochem Biophys Res Commun 98: 952–960

Gorbalenya AE, Blinov VM, Donchenko AP (1986) Poliovirus-encoded proteinase 3C: a possible evolutionary link between cellular serine and cysteine proteinase families. FEBS Lett 194: 253–257

Gorbalenya AE, Koonin EV, Blinov VM, Donchenko AP (1988) Sobemovirus genome appears to encode a serine protease related to cysteine proteases of picornaviruses. FEBS Lett 236: 287–290

Gorbalenya AE, Donchenko AP, Blinov VM, Koonin EV (1989) Cysteine proteases of positive strand RNA viruses and chymotrypsin-like serine proteases: a distinct protein superfamily with a common structural fold. FEBS Lett 243: 103–114

Granboulan N, Girard M (1969) Molecular weight of poliovirus ribonucleic acid. J Virol 4: 475–479

Greve JM, Davis G, Meyer AM, Forte CP, Yost SC, Marlor CW, Kamarck ME, McClelland A (1989) The major human rhinovirus receptor is ICAM-1. Cell 56: 839–847

Grubman MJ, Baxt B (1982) Translation of foot-and mouth disease virion RNA and processing of the primary cleavage products in a rabbit reticulocyte lysate. Virology 116: 19–30

Grubman MJ, Robertson BH, Morgan DO, Moore DM, Dowbenko D (1984) Biochemical map of polypeptides specified by foot-and-mouth disease virus. J Virol 50: 579–586

Hanecak R, Semler BL, Anderson CW, Wimmer E (1982) Proteolytic processing of poliovirus polypeptides: antibodies to a polypeptide P3-7c inhibit cleavage at glutamine-glycine pairs. Proc Natl Acad Sci USA 79: 3973–3977

Hanecak R, Semler BL, Ariga H, Anderson CW, Wimmer E (1984) Expression of a cloned gene segment of poliovirus in E. coli: evidence for autocatalytic production of the viral proteinase. Cell 37: 1063–1073

Higaki JN, Gibson BW, Craik CS (1987) Evolution of catalysis in the serine proteases. Cold Spring Harb Symp Quant Biol 52: 615–621

Hogle JM, Chow M, Filman DJ (1985) The three dimensional structure of poliovirus at 2.9 Å resolution. Science 229: 1358–1365

Holland JJ, Kiehn ED (1968) Specific cleavage of viral proteins as steps in the synthesis and maturation of enteroviruses. Proc Natl Acad Sci USA 60: 1015–1022

Iizuka N, Kuge S, Nomoto A (1987) Complete nucleotide sequence of the genome of coxsackievirus B1. Virology 156: 64–73

Ivanoff LA, Towatari T, Ray J, Korant BD, Petteway SR (1986) Expression and site-specific mutagenesis of the poliovirus 3C protease in Escherichia coli. Proc Natl Acad Sci USA 83: 5392–5396

Jackson RJ (1986) A detailed kinetic analysis of the in vitro synthesis and processing of encephalomyocarditis virus products. Virology 149: 114–127

Jackson RJ (1989) Comparison of encephalomyocarditis virus and poliovirus with respect to translation initiation and processing in vitro. In: Semler BL, Ehrenfeld E (eds) Molecular aspects of picornavirus infection and detection. American Society for Microbiology, Washington DC, pp 51–71

Jacobson MF, Baltimore D (1968a) Morphogenesis of poliovirus I: Association of the viral RNA with coat protein. J Mol Biol 33: 369–378

Jacobson MF, Baltimore D (1968b) Polypeptide cleavages in the formation of poliovirus proteins. Proc Natl Acad Sci USA 61: 77–84

Jacobson MF, Asso J, Baltimore D (1970) Further evidence on the formation of poliovirus proteins. J Mol Biol 49: 657–669

Jenkins O, Booth JD, Minor PD, Almond JW (1987) The complete nucleotide sequence of coxsackievirus B4 and its comparison to other members of the picornaviridae. J Gen Virol 68: 1835–1848

Johnson VH, Semler BL (1988) Defined recombinants of poliovirus and coxsackievirus: sequence-specific deletions and functional substitutions in the 5'-noncoding regions of viral RNAs. Virology 162: 47–57

Jore J, De Geus B, Jackson RJ, Pouwels PH, Enger-Valk BE (1988) Poliovirus protein 3CD is the active protease for processing of the precursor protein P_1 in vitro. J Gen Virol 69:1627–1636

Kalderon D, Oostra BA, Ely BK, Smith AE (1982) Deletion loop mutagenesis: a novel method for the construction of point mutants using deletion mutants. Nucleic Acids Res 10: 5161–5171

Kean KM, Agut H, Fichot O, Wimer E, Girard M (1988) A poliovirus mutant defective for self cleavage at the COOH-terminus of the 3C protease exhibits secondary processing defects. Virology 163: 330–340

Kew OM, Pallansch MA, Omilianowski DR, Rueckert RR (1980) Changes in three of the four coat proteins of oral polio vaccine strain derived from type 1 poliovirus. J Virol 33: 256–263

Kiehn ED, Holland JJ (1970) Synthesis of enterovirus polypeptides in mammalian cells. J Virol 5: 358–367

King AMQ, Sangar DV, Harris TJR, Brown F (1980) Heterogeneity of the genome-linked protein of foot-and-mouth disease virus. J Virol 34: 627–634

Kitamura N, Semler BL, Rothberg PG, Larsen GR, Adler CJ, Dorner AJ, Emini EA, Hanecak R, Lee JJ, van der Werf S, Anderson CW, Wimmer E (1981) Primary structure, gene organization, and polypeptide expression of poliovirus RNA. Nature 291: 547–553

Klump W, Marquardt O, Hofschneider PH (1984) Biologically active protease of foot and mouth disease virus is expressed from cloned viral cDNA in Escherichia coli. Proc Natl Acad Sci USA 81: 3351–3355

König H, Rosenwirth B (1988) Purification and partial characterization of poliovirus protease 2A by means of a functional assay. J Virol 62: 1243–1250

Korant BD (1972) Cleavage of viral precursor proteins in vivo and in vitro. J Virol 10: 751–759

Korant BD (1973) Cleavage of poliovirus-specific polypeptide aggregates. J Virol 12: 556–563

Korant B, Chow N, Lively M, Powers J (1979) Virus-specified protease in poliovirus-infected HeLa cells. Proc Natl Acad Sci USA 76: 2992–2995

Korant BD, Brzin J, Turk V (1985) Cystatin, a protein inhibitor of cysteine proteases alters viral protein cleavages in infected human cells. Biochem Biophys Res Commun 127: 1072–1076

Kräusslich HG, Wimmer E (1988) Viral Proteinases. Ann Rev Biochem 57: 701–754

Kräusslich HG, Nicklin MJH, Toyoda H, Etchison D, Wimmer E (1987) Poliovirus proteinase 2A induces cleavage of eucaryotic initiation factor 4F polypeptide p220. J Virol 61: 2711–2718

Kräusslich HG, Nicklin MJH, Lee C, Wimmer E (1988) Polyprotein processing in picornavirus replication. Biochimie 70: 119–130

Kuhn RJ, Tada H, Ypma-Wong MF, Dunn JJ, Semler BL, Wimmer E (1988a) Construction of a "mutagenesis cartridge" for poliovirus genome-linked viral protein: isolation and characterization of viable and non-viable mutants. Proc Natl Acad Sci USA 85: 519–523

Kuhn RJ, Tada H, Ypma-Wong MF, Semler BL, Wimmer E (1988b) Mutational analysis of the genome-linked protein VPg of poliovirus. J Virol 62: 4207–4215

Larsen GR, Anderson CW, Dorner AJ, Semler BL, Wimmer E (1982) Cleavage sites within the poliovirus capsid protein precursors. J Virol 41: 340–344

Lawrence C, Thach RE (1975) Identification of a viral protein involved in post-translational maturation of the encephalomyocarditis virus capsid precursor. J Virol 15: 918–928

Lee C, Wimmer E (1988) Proteolytic processing of poliovirus polyprotein: elimination of 2A^{pro}-mediated alternative cleavage of polypeptide 3CD by in vitro mutagenesis. Virology 166: 405–414

Lee KAW, Edery I, Hanecak R, Wimmer E, Sonenberg N (1985) Poliovirus protease 3C (P3-7C) does not cleave P220 of the eukaryotic mRNA cap-binding protein complex. J Virol 55: 489–493

Lee YF, Nomoto A, Detjen BM, Wimmer E (1977) A protein covalently linked to poliovirus genome RNA. Proc Natl Acad Sci USA 74: 59–63

Libby RT, Cosman D, Cooney MK, Merriam JE, March CJ, Hopp TP (1988) Human rhinovirus 3C protease: cloning and expression of an active form in *Escherichia coli*. Biochemistry 27: 6262–6268

Lindberg AM, Stalhandske POK, Pettersson U (1987) Genome of coxsackievirus B3. Virology 156: 50–63

Lipton HL, Rozhon EJ, Black D (1984) Theiler's virus-specified polypeptides made in BHK-21 cells. J Gen Virol 65: 1095–1100

Lloyd RE, Etchison D, Ehrenfeld E (1985) Poliovirus protease does not mediate cleavage of the 220,000-Da component of the cap binding protein complex. Proc Natl Acad Sci USA 82 2723–2727

Lloyd RE, Toyoda H, Etchison D, Wimmer E, Ehrenfeld E (1986) Cleavage of the cap binding protein complex polypeptide p220 is not effected by the second poliovirus protease 2A. Virology 150: 299–303

Lloyd RE, Jense HG, Ehrenfeld E (1987) Restriction of translation of capped mRNA in vitro as a model for poliovirus-induced inhibition of host cell protein synthesis: relationship to p220 cleavage. J Virol 61: 2480–2488

Lloyd RE, Grubman MJ, Ehrenfeld E (1988) Relationship of p220 cleavage during picornavirus infection to 2A proteinase sequences. J Virol 62: 4216–4223

Loesch WT, Arlinghaus RB (1974) Polypeptides associated with the 250S mengovirus-induced RNA polymerase structure. Arch Gesamte Virusforsch 46: 253–268

Luo M, Vriend G, Kamer G, Minor I, Arnold E, Rossmann MG, Boege U, Scraba DG, Duke GM, Palmenberg AC (1987) The atomic structure of mengo virus at 3.0 Å resolution. Science 235: 182–191

Maizel JV (1963) Evidence for multiple components in the structural protein of type 1 poliovirus. Biochem Biophys Res Commun 13: 483–489

Maizel JV, Summers DF (1968) Evidence for differences in size and composition of the poliovirus-specific polypeptides in infected HeLa cells. Virology 36: 45–54

Maizel JV, Phillips BA, Summers DF (1967) Composition of artificially produced and naturally occurring empty capsids of poliovirus type 1. Virology 32: 692–699

McClean C, Rueckert RR (1973) Picornaviral gene order: comparison of a rhinovirus with a cardiovirus. J Virol 11: 341–344

Mendelsohn CL, Wimmer E, Racaniello VR (1989) Cellular receptor for poliovirus: molecular cloning, nucleotide sequence, and expression of a new member of the immunoglobulin superfamily. Cell 56: 855–865

Morrow CD, Hocko J, Navab M, Dasgupta A (1984) ATP is required for initiation of poliovirus RNA synthesis in vitro: demonstration of tyrosine-phosphate linkage between in vitro-synthesized RNA and genome-linked protein. J Virol 50: 515–523

Mosenskis J, Daniels-McQueen S, Janovec S, Duncan R, Hershey JWB, Grifo JA, Merrick WC, Thach RE (1985) Shutoff of host translation by encephalomyocarditis virus infection does not involve cleavage of the eucaryotic initiation factor 4F polypeptide that accompanies poliovirus infection. J Virol 54: 643–645

Neurath H (1984) Evolution of proteolytic enzymes. Science 224: 350–357

Nicklin MJH, Toyoda H, Murray MG, Wimmer E (1986) Proteolytic processing in the replication of polio and related viruses. Biotechnology 4: 36–42

Nicklin MJH, Kräusslich HG, Toyoda H, Dunn JJ, Wimmer E (1987) Poliovirus polypeptide precursors: expression in vitro and processing by 3C and 2A proteinases. Proc Natl Acad Sci USA 84: 4002–4006

Nicklin MJH, Harris KS, Pallai PV, Wimmer E (1988) Poliovirus proteinase 3C: large scale expression, purification, and specific cleavage activity on natural and synthetic substrates in vitro. J Virol 62: 4586–4593

Nomoto A, Detjen BM, Pozzatti R, Wimmer E (1977) Location of the polio genome protein in viral RNAs and its implication for RNA synthesis. Nature 268: 208–213

Ohara Y, Stein S, Fu J, Stillman L, Klaman L, Roos RP (1988) Molecular cloning and sequence determination of DA strain of Theiler's murine encephalomyelitis viruses. Virology 164: 245–255

Pallai PV, Burkhardt F, Skoog M, Schreiner K, Baxt P, Cohen KA, Hansen G, Palladino DEH, Harris KS, Nicklin MJ, Wimmer E (1989) Cleavage of synthetic peptides by purified poliovirus 3C proteinase. J Biol Chem 264: 9738–9741

Pallansch MA, Kew OM, Semler BL, Omilianowski DR, Anderson CW, Wimmer E, Rueckert RR (1984) Protein processing map of poliovirus. J Virol 49: 873–880

Palmenberg AC, Rueckert RR (1982) Evidence for intramolecular self-cleavage of picornaviral replicase precursors. J Virol 41: 244–249

Palmenberg AC, Pallansch MA, Rueckert RR (1979) Protease required for processing picornaviral coat protein resides in the viral replicase gene. J Virol 32: 770–778

Palmenberg AC, Kirby EM, Janda MR, Duke GM, Potratz KF, Collett MS (1984) The nucleotide sequence and deduced amino acid sequences of the encephalomyocarditis viral polyprotein coding region. Nucleic Acids Res 12: 2969–2985

Parks GD, Palmenberg A (1987) Site specific mutations at a picornavirus VP3/VP1 cleavage site disrupt in vitro processing and assembly of capsid precursors. J Virol 61: 3680–3687

Parks GD, Duke GM, Palmenberg AC (1986) Encephalomyocarditis virus 3C protease: efficient cell-free expression from clones which link viral 5′ noncoding sequences to the P3 region. J Virol 60: 376–384

Parks GD, Baker JC, Palmenberg AC (1989) Proteolytic cleavage of encephalomyocarditis virus capsid region substrates by precursors to the 3C enzyme. J Virol 63: 1054–1058

Paucha E, Seehafer J, Colter JG (1974) Synthesis of viral-specific polypeptides in mengo virus-infected L cells: evidence for asymetric translation of the viral genome. Virology 61: 315–326

Pelham HRB (1978) Translation of encephalomyocarditis virus RNA in vitro yields an active proteolytic processing enzyme. Eur J Biochem 85: 425–462

Pelham HRB, Jackson RJ (1976) An efficient mRNA-dependent translation system from reticulocyte lysates. Eur J Biochem 67: 247–256

Penman S, Summers DF (1965) Effects on host cell metabolism following synchronous infection with poliovirus. Virology 27: 614–620

Penman S, Becker Y, Darnell JE (1964) A cytoplasmic structure involved in the synthesis and assembly of poliovirus components. J Mol Biol 8: 541–555

Pevear DC, Calenoff M, Rozhon E, Lipton HL (1987) Analysis of the complete nucleotide sequence of the picornavirus Theiler's murine encephalomyelitis virus indicates that it is closely related to cardioviruses. J Virol 61: 1507–1516

Plotch SJ, Palant O, Gluzman Y (1989) Purification and properties of poliovirus RNA polymerase expressed in Escherchia coli. J Virol 63: 216–225

Polgár L, Halász P (1982) Current problems in mechanistic studies of serine and cysteine proteinases. Biochem J 207: 1–10

Putnak JR, Phillips BA (1981) Picornaviral structure and assembly. Microbiol Rev 45: 287–315

Racaniello VR, Baltimore D (1981) Molecular cloning of poliovirus and determination of the complete nucleotide sequence of the viral genome. Proc Natl Acad Sci USA 78: 4887–4891

Rekosh D (1972) Gene order of the poliovirus capsid proteins. J Virol 9: 479–487

Robertson BH, Grubman MJ, Weddel GN, Moore DM, Welsh JD, Fischer T, Dowbenko DJ, Yansura DG, Small B, Kleid DG (1985) Nucleotide and amino acid sequence coding for polypeptides of foot-and-mouth disease virus type A12. J Virol 54: 651–660

Roos RP, Kong W, Semler BL (1989a) Polyprotein processing of Theiler's murine encephalomyelitis virus. J Virol 63: 5344–5353

Roos RP, Stein S, Ohara Y, Fu J, Semler BL (1989b) Infectious cDNA clones of DA strain of Theiler's murine encephalomyelitis virus. J Virol 63: 5492–5496

Rossmann MG, Arnold E, Erickson JW, Frankenberger EA, Griffith JP, Hecht H, Johnson JE, Kamer G, Luo M, Mosser AG, Rueckert RR, Sherry B, Vriend G (1985) Structure of a human common cold virus and functional relationship to other picornaviruses. Nature 317: 145–153

Rothberg PG, Harris TJR, Nomoto A, Wimmer E (1978) O[4]-(5′-uridylyl) tyrosine is the bond between the genome-linked protein and the RNA of poliovirus. Proc Natl Acad Sci USA 75: 4868–4872

Rueckert RR (1985) Picornaviruses and their replication. In: Fields BN (ed) Virology. Raven, New York, pp 705–738

Rueckert RR, Wimmer E (1984) Systematic nomenclature of picornavirus proteins. J Virol 50: 957–959

Rueckert RR, Matthews TJ, Kew OM, Pallansch M, McLean C, Omilianowski D (1979) Synthesis and processing of picornaviral polyproteins. In: Perez-Bercoff R (ed) The molecular biology of picornaviruses. Plenum, New York, pp 113–125

Salzman NP, Lockhart RZ, Sebring ED (1959) Alterations in HeLa cell metabolism resulting from poliovirus infection. Virology 9: 244–259

Sangar DV, Black DN, Rowlands DJ, Brown F (1977) Biochemical mapping of the foot-and-mouth disease virus genome. J Gen Virol 35: 281–297

Sangar DV, Newton, SE, Rowlands DJ, Clarke BE (1987) All foot and mouth disease virus serotypes initiate protein synthesis at two separate AUGs. Nucleic Acids Res 15: 3305–3315

Sangar DV, Clark RP, Carroll AR, Rowlands DJ, Clarke BE (1988) Modification of the leader protein (Lb) of foot-and-mouth disease virus. J Gen Virol 69: 2327–2333

Schaffer FL (1962) Physical and chemical properties and infectivity of RNA from animal viruses. Cold Spring Harb Symp Quant Biol 27: 89–99

Schaffer FL, Schwerdt CE (1959) Purification and properties of poliovirus. Adv Virus Res 6: 159–204

Semler BL, Hanecak R, Anderson CW, Wimmer E (1981a) Cleavage sites in the polypeptide precursors of poliovirus protein P2-X. Virology 114: 589–594

Semler BL, Anderson CW, Kitamura N, Rothberg PG, Wishart WL, Wimmer E (1981b) Poliovirus replication proteins: RNA sequence encoding P3-1b and the sites of proteolytic processing. Proc Natl Acad Sci USA 78: 3464–3468

Semler BL, Anderson CW, Hanecak R, Dorner LF, Wimmer E (1982) A membrane-associated precursor to poliovirus VPg identified by immunoprecipitation with antibodies directed against a synthetic heptapeptide. Cell 28: 405–412

Semler BL, Hanecak R, Dorner LF, Anderson CW, Wimmer E (1983) Poliovirus RNA synthesis in vitro: structural elements and antibody inhibition. Virology 126: 624–634

Semler BL, Johnson VH, Dewalt PG, Ypma-Wong MF (1987) Site specific mutagenesis of cDNA clones expressing a poliovirus proteinase. J Cell Biochem 33: 39–51

Shih DS, Shih CT, Zimmern D, Rueckert RR, Kaesberg P (1979) Translation of encephalomyocarditis virus RNA in reticulocyte lysates: kinetic analysis of the formation of virion proteins and a protein required for processing. J Virol 30: 472–480

Skern T, Sommergruber W, Blaas D, Gruendler P, Fraundorfer F, Pieler C, Fogy I, Kuechler E (1985) Human rhinovirus 2: complete nucleotide sequence and proteolytic processing signals in the capsid protein region. Nucleic Acids Res 13: 2111–2126

Sonenberg N (1987) Regulation of translation by poliovirus. Adv Virus Res 33: 175–204

Stanway G, Hughes PJ, Mountford RC, Minor PD, Almond JW (1984a) The complete sequence of a common cold virus: human rhinovirus 14. Nucleic Acids Res 12: 7859–7874

Stanway G, Hughes PJ, Mountford RC, Reeve P, Minor PD, Schild GC, Almond JW (1984b) Comparison of the complete nucleotide sequence of the genomes of the neurovirulent poliovirus P3/Leon/37 and its attenuated Sabin vaccine derivative P3/Leon12a$_1$b. Proc Natl Acad Sci USA 81: 1539–1543

Staunton DE, Merluzzi VJ, Rothlein R, Barton R, Marlin S, Springer TA (1989) A cell adhesion molecule, ICAM-1 is the major surface receptor for rhinoviruses. Cell 56: 849–853

Strebel K, Beck E (1986) A second protease of foot-and-mouth disease virus. J Virol 58: 893–899

Summers DF, Levintow L (1965) Constitution and function of polyribosomes in poliovirus infected HeLa cells. Virology 27: 44–53

Summers DF, Maizel JV (1968) Evidence for large precursor proteins in poliovirus synthesis. Proc Natl Acad Sci USA 59: 966–971

Summers DF, Maizel JV (1971) Determination of the gene sequence of poliovirus with pactamycin. Proc Natl Acad Sci USA 68: 2852–2856

Summers DF, Maizel JV, Darnell JE (1965) Evidence for virus-specific noncapsid proteins in poliovirus-infected HeLa cells. Proc Natl Acad Sci USA 54: 505–513

Summers DF, Shaw EN, Stewart ML, Maizel JV (1972) Inhibition of cleavage of large poliovirus-specific precursor proteins in infected HeLa cells by inhibitors of proteolytic enzymes. J Virol 10: 880–884

Svitkin YV, Gorbalenya AE, Kazachkov YA, Agol VI (1979) Encephalomyocarditis virus-specific polypeptide p22 possessing proteolytic activity. FEBS Lett 108: 6–9

Taber R, Rekosh D, Baltimore D (1971) Effect of pactamycin on synthesis of poliovirus proteins: a method for genetic mapping. J Virol 8: 395–401

Takeda N, Kuhn RJ Yang C-F, Takegami T, Wimmer E (1986) Initiation of poliovirus plus-strand RNA synthesis in a membrane complex of infected HeLa cells. J Virol 60: 43–53

Takegami T, Kuhn RJ, Anderson CW, Wimmer E (1983) Membrane-dependent uridylylation of the genome-linked protein VPg of poliovirus. Proc Natl Acad Sci USA 80: 7447–7451

Ticehurst J, Cohen JI, Feinstone SM, Purcell RH, Jansen RW, Lemon SM (1989) Replication of hepatitis A virus: new ideas from studies with cloned cDNA. In: Semler BL, Ehrenfeld E (eds) Molecular aspects of picornavirus infection and detection. American Society for Microbiology, Washington DC, pp 27–50

Toyoda H, Nicklin MJH, Murray MG, Anderson CW, Dunn JJ, Studier FW, Wimmer E (1986) A second virus-encoded proteinase involved in proteolytic processing of poliovirus polyprotein. Cell 45: 761–770

Toyoda H, Yang C, Takeda N, Nomoto A, Wimmer E (1987) Analysis of RNA synthesis of type 1 poliovirus by using an in vitro molecular genetic approach. J Virol 61: 2816–2822

Tracy S, Liu H-L, Chapman N (1985) Coxsackievirus B3: primary structure of the 5' non-coding region and capsid protein-coding regions of the genome. Virus Res 3: 263–270

Vakharia VN, Devaney MA, Moore DM, Dunn JJ, Grubman MJ (1987) Proteolytic processing of foot-and-mouth disease virus polyproteins expressed in a cell-free system from clone-derived transcripts. J Virol 61: 3199–3207

Vartapetian AB, Drygin YF, Chumakov KM, Bogdanov AA (1980) The structure of the covalent linkage between proteins and RNA in encephalomyocarditis virus. Nucleic Acids Res 8: 3729–3741

Villa-Komaroff L, Guttman N, Baltimore D, Lodish HF (1975) Complete translation of poliovirus RNA in a eukaryotic cell-free system. Proc Natl Acad Sci USA 72: 4157–4161

Werner G, Rosenwirth B, Bauer E, Seifert J-M, Werner, F-J, Besemer J (1986) Molecular cloning and sequence determination of the genomic regions encoding protease and genome-linked protein of three picornaviruses. J Virol 57: 1084–1093

Wiegers KJ, Dernick R (1981a) Poliovirus-specific polypeptides in infected HeLa cells analysed by isoelectric focusing and 2D-analysis. J Gen Virol 52: 61–69

Wiegers KJ, Dernick R (1981b) Peptide maps of labelled poliovirus proteins after two-dimensional analysis by limited proteolysis and electrophoresis in sodium dodecyl sulfate. Electrophoresis 2: 98–103

Yogo Y, Wimmer E (1972) Polyadenylic acid at the 3' terminus of poliovirus RNA. Proc Natl Acad Sci USA 69: 1877–1882

Young DC, Tuschall DM, Flanegan JB (1985) Poliovirus RNA-dependent RNA polymerase and host cell protein synthesize product RNA twice the size of poliovirion RNA in vitro. J Virol 54: 256–264

Ypma-Wong MF, Semler BL (1987a) In vitro molecular genetics as a tool for determining the differential cleavage specificities of the poliovirus 3C proteinase. Nucleic Acids Res 15: 2069–2088

Ypma-Wong MF, Semler BL (1987b) Processing determinants required for in vitro cleavage of the poliovirus P1 precursor to capsid proteins. J Virol 61: 3181–3189

Ypma-Wong MF, Filman DJ, Hogle JM, Semler BL (1988a) Structural domains of the poliovirus polyprotein are major determinants for proteolytic cleavage at gln-gly pairs. J Biol Chem 263: 17846–17856

Ypma-Wong MF, Dewalt PG, Johnson VH, Lamb JG, Semler BL (1988b) Protein 3CD is the major poliovirus proteinase responsible for cleavage of the P1 capsid precursor. Virology 166: 265–270

Zimmerman EF, Heeter M, Darnell JE (1963) RNA synthesis in poliovirus-infected cells. Virology 19: 400–408

Ziola BR, Scraba DG (1976) Structure of the Mengo virion IV. Amino- and carboxyl-terminal analyses of the major capsid polypeptides. Virology 71: 111–121

Poliovirus RNA Replication*

O. C. Richards and E. Ehrenfeld

Departments of Biochemistry and Cellular, Viral, Molecular Biology, University of Utah School of Medicine, Salt Lake City, Utah 84132, USA
* Work on poliovirus RNA replication in the authors' laboratories was supported by a grant from the US Public Health Service (AI 17386)

1 Introduction

The biosynthesis of RNA directed by an RNA template is a reaction that is unique to RNA viruses. Although studies of poliovirus RNA synthesis have been conducted in a somewhat intermittent fashion during the past 25 years in several different laboratories, no clear picture has yet emerged regarding the biochemistry of RNA replication for this or any other RNA virus. Upon entry into the cell, the positive strand, infecting RNA genome directs the synthesis of viral proteins which are required for replication of the RNA. The replication process involves first the synthesis of a negative strand RNA molecule; subsequent transcription of this negative strand produces new copies of the positive strand RNA. Historically, the experimental approach initially utilized to analyze the poliovirus RNA replication reaction was enzymological; efforts were made to isolate and purify an RNA-dependent RNA polymerase activity from virus-infected cells. Indeed, at that time, the only tools available for RNA replication studies were biochemical. The biochemistry, however, proved difficult. RNA replication was found to occur in intracellular structures that are tightly associated with or in membranes, and these proved intractable to purification and dissection. Disruption of the membrane structure in order to isolate template or enzyme components often appeared to alter their properties and/or structures. Thus, the initial approach yielded little information about the mechanism of RNA replication, and it has been only quite recently that alternative approaches have been applied.

By comparison, the RNA polymerases that catalyze DNA-dependent RNA synthesis have been much better characterized. Such enzymes have been purified from numerous prokaryotic and eukaryotic sources, and their polypeptide compositions, template recognition (promoter binding), and catalytic activities are under continued productive investigation. DNA-binding enzymes must recognize a specific sequence of duplex DNA (promoter) and induce unwinding of the duplex to allow initiation and subsequent movement of a transcription complex. In contrast, the polymerases that utilize single-stranded RNA templates have the unique problem of starting *de novo* at an end, with no obvious way to anchor the enzyme or bind the first nucleotide. A similar problem is faced by reverse transcriptases, which synthesize DNA from an RNA template. These enzymes have been shown to rely on a striking diversity of initiation and priming mechanisms: intact or fragmented tRNA molecules, oligoribonucleotides generated by RNase H activity, and proteins all have been identified as primers for reverse transcriptases in different biological systems. Most recently, priming of DNA synthesis by the 2' hydroxyl of an internal ribonucleotide to generate a branched template-product structure has been elucidated in bacteria. In most cases, priming occurs at internal locations on the templates, but examples of initiation at a strand terminus by a reverse transcriptase are also known.

The mechanism utilized to initiate poliovirus RNA synthesis has not been elucidated, but the variety of reactions in which RNA molecules have recently been demonstrated to participate has enlarged the scope of possible scenarios. This review will describe the information currently available about poliovirus RNA replication, and will attempt to present current models in the hopes that we may focus some of the questions that should be addressed by new experimental approaches.

2 RNA Synthesis in Infected Cells

2.1 General Description of Viral RNA Synthesis

Infection of cultured cells with poliovirus results in a brief latent period, followed by the appearance of viral RNA and progeny virus. Initial studies of the infection cycle monitored the increase in infectious RNA with time and the incorporation of labeled uridine into viral RNA in the presence of actinomycin D (DARNELL et al. 1961; SCHARFF et al. 1963). Detectable viral RNA appeared at about 2 h postinfection and rapidly rose thereafter, preceding the appearance of virus by about 30 min. The synthesis of viral RNA was observed to occur solely in the cytoplasm of infected cells (FRANKLIN and BALTIMORE 1962). Enucleated cells could fully support poliovirus replication, confirming that RNA replication required no nuclear component (CROCKER et al. 1964). Newly synthesized RNA was packaged into virus particles within 5 min (BALTIMORE et al. 1966), indicating that the pool of viral RNA in the cell was small. At the time of maximal rates of RNA synthesis (3–4 h postinfection), it took between 3/4 and 1 min to complete the synthesis of a viral RNA molecule, and production approached 2000–3000 molecules per cell per minute (DARNELL et al. 1967). At early times ($1\frac{1}{2}$–3 h postinfection), viral RNA was largely localized in polysomes, whereas RNA synthesized subsequently was predominantly assembled into viral particles.

2.2 Viral RNAs in Infected Cells

The first viral RNA species detectable after pulse-labeling infected cells is a heterogeneous, multistranded structure called replicative intermediate (RI; NOBLE and LEVINTOW 1970). It is maintained at a low steady-state concentration throughout the infectious cycle. The predominant species of viral RNA that accumulates in infected cells is single-stranded RNA (ssRNA), identical to that which is found in virions. No subgenomic RNAs have been isolated during lytic infection of cells with laboratory strains of virus. Viral double-stranded RNA (dsRNA) is continuously formed at a relatively slow rate. Each of these three species of viral RNA is described in some detail, below.

2.2.1 Single-Stranded RNA

The major RNA species found in infected cells is ssRNA of positive polarity. It is packaged into virion particles which remain in the cytoplasm until cell lysis, and it also serves as the only viral mRNA bound to large polysome structures. It has a sedimentation coefficient of 35S in sucrose gradients in 0.1 M NaCl (ZIMMERMAN et al. 1963). When the purified RNA is artificially introduced into cells, it is sufficient to generate viral progeny (ALEXANDER et al. 1958; HOLLAND et al. 1959). The 3' end of the RNA consists of a variable stretch of approximately 90 polyadenylate (poly(A)) residues (YOGO and WIMMER 1972; SPECTOR and BALTIMORE 1975a); and a small protein, VPg (3B), is covalently attached at the 5' terminus (FLANEGAN et al. 1977; LEE et al. 1977). The specific linkage of VPg to the polynucleotide chain is a phosphoester bond between a tyrosine residue in VPg and the 5' phosphate of the terminal uridine nucleotide (ROTHBERG et al. 1978; AMBROS and BALTIMORE 1978). VPg is apparently removed from those molecules functioning as mRNA (NOMOTO et al. 1976; HEWLETT et al. 1976; FERNANDEZ-MUNOZ and DARNELL 1976). The presence of VPg on the RNA is not required for infectivity, as virion RNA and mRNA are equally infectious (NOMOTO et al. 1976, 1977a). The genome of poliovirus, Mahoney strain type 1, is 7441 nucleotides in length, exclusive of its poly(A) tract, and it has been completely sequenced (KITAMURA et al.1981; RACANIELLO and BALTIMORE 1981). It contains a 5' noncoding end of 742 nucleotides and a 3' noncoding sequence of 71 nucleotides, just upstream of the poly(A) tail. Complete sequences have also been determined for the Sabin strains of each of the three serotypes (NOMOTO et al. 1982; TOYODA et al. 1984), as well as the virulent parent strains from which they were derived (KITAMURA et al. 1981; RACANIELLO and BALTIMORE 1981; LA MONICA et al. 1986; STANWAY et al. 1983). The mature, viral-encoded polypeptides, formed during a poliovirus infection have been delineated and will not be described here (RUECKERT and WIMMER 1984; see Chapter 3 by LAWSON and SEMLER, this volume).

2.2.2 Replicative Intermediates

RI have been described as a class of molecules which have a genome-length strand and varying numbers of complementary strands with lengths ≤ genomic length (BALTIMORE and GIRARD 1966; GIRARD et al. 1967; GIRARD 1969; BISHOP and KOCH 1969; BISHOP et al. 1969; SAVAGE et al. 1971). The structures are partially RNase resistant, as isolated from infected cells, and the RNase-resistant product is a double-stranded "core" molecule. In sucrose gradients, RI structures sediment as a heterogeneous population of molecules of 20–70S; in equilibrium gradients they band at densities intermediate between ssRNA and dsRNA. As a class they have between 4 and 8 nascent strands (branches) per molecules. After removal of the single-stranded, nascent "tails" with RNase, 90% of the remaining genome-length backbone strands are of negative polarity, and 10% are plus strands; furthermore, the RNase-resistant product contains 10% of the original infectivity. These findings suggest that this population of RNA molecules may be subdivided into a predominant class of plus-strand RI (having nascent plus strands) and a minor class of minus-strand RI (having nascent minus strands).

Classically, as isolated by phenol extraction from poliovirus-infected cells, RI appears in the electron microscope as a genome-length double-stranded backbone with single strands of different lengths emanating from the backbone (SAVAGE et al. 1971; RICHARDS et al. 1984). More recent studies using covalent cross-linking and denaturation techniques support the view that RI have predominantly single-stranded backbones in vivo with nascent strands hydrogen-bonded to their template strand only transiently at replication forks (RICHARDS et al. 1984). All strands (template and nascent) are already covalently attached to VPg by the time that they can be detected (when nascent strands have achieved a length of 300 nucleotides; NOMOTO et al. 1977b; PETTERSSON et al. 1978). Thus, VPg must be added to nascent strands at the time of initiation of strand synthesis or soon thereafter. The 3′ terminus of the plus strand of RI contains poly(A), just as virion RNA does, and the minus strand contains poly(U) (YOGO and WIMMER 1975; SPECTOR and BALTIMORE 1975a, b).

2.2.3 Double-Stranded RNA

The first description of viral dsRNA in picornavirus-infected cells was reported by MONTAGNIER and SANDERS (1963). This molecule was shown to contain one genome length plus strand and one genome length minus strand, to sediment at 18–20S in sucrose gradients, and to be soluble in 1 M NaCl and display a thermal transition characteristic of a dsRNA molecule (BALTIMORE 1966; BISHOP and KOCH 1967). Studies of the termini of dsRNA molecules revealed that one end of both plus and minus strands was heteropolymeric whereas the other end contained a poly(A) stretch (3′ end of plus strands) or a poly(U) stretch (5′ end of minus strands), suggesting that each homopolymeric stretch was coded by the complementary strand (YOGO and WIMMER 1973; SPECTOR and BALTIMORE 1975b; LARSEN et al. 1980). VPg was found at the 5′ terminus of both strands (WU et al. 1978; LARSEN et al. 1980). Some molecules of dsRNA had extra adenylate residues on the 3′ ends of the minus strands (RICHARDS and EHRENFELD 1980) and dsRNA molecules, as a class, have poly(A) tracts which considerably exceed the complementary poly(U) stretch on minus strands (LARSEN et al. 1980). Each of these observations suggests a noncoded terminal addition of nucleotides to the 3′ ends of strands and/or a polymerase slippage phenomenon. Late in poliovirus infection some minus strands appear to lack VPg (RICHARDS et al. 1979). Perhaps related to this observation, YOUNG et al. (1985) and RICHARDS et al. (1987) report that some dsRNA are "hairpin" structures, covalently linked at one end which, upon denaturation, migrate in denaturing gels as polynucleotide chains that are twice genome length. A long-standing question remains whether dsRNA molecules are intermediates in poliovirus replication or are dead-end products, unable to support further replication. MEYER et al. (1978) found dsRNA in isolated poliovirus replication complexes, but this does not necessarily implicate them in the replicative process. Interestingly, all minus strands are associated with RI or dsRNA in vivo; no free minus strands are detectable (ROY and BISHOP 1970).

2.3 Association of RNA Replication Complexes with Membranes

The first evidence for a poliovirus RNA polymerase activity in infected cells was reported by BALTIMORE et al. (1963). The activity was found in the microsomal fraction of infected cells, and the kinetics of its appearance closely paralleled virus production. Velocity sedimentation in sucrose gradients of cytoplasmic extracts yielded polymerase activity in a large complex which was associated with intracellular membranes. The complex was pulse-labeled with [^{3}H]uridine, and was isolated with the smooth membrane fraction from infected cells (CALIGUIRI and TAMM 1970a; TAKEGAMI et al. 1983a). A marked proliferation of intracellular smooth membranes occurs after infection (MOSSER et al. 1972); the structure of these membranes and the nature of their interaction with the viral replication complex have been visualized by electron microscopy (BIENZ et al. 1987), but have not been studied biochemically. Treatment of cytoplasmic extracts with deoxycholate released from the membranes a replication complex which sedimented at 250S (PENMAN et al. 1964; BALTIMORE 1964; GIRARD et al. 1967).

3 The Replication Complex

As will be described below, the crude replication complexes isolated from infected cells are active in polio RNA synthesis, contain endogenous RNA templates, viral proteins 3D, 2C, and precursor forms of VPg, and perhaps additional polypeptides and factors essential for RNA synthesis. Activities for the direct addition of two nucleotides to a VPg-containing protein, and for RNA chain elongation, can be measured in these complexes. The various components are organized in a complex, membranous structure, whose integrity appears to be required for proper function. Since the stoichiometries of the various reactions have not been established in the crude complexes, and the system has not proven amenable to simple biochemical dissection, the precise contributions of each of these observed reactions to the actual polio RNA replication scheme remains uncertain.

3.1 Viral Components of the Crude Complexes

The large, membrane-associated replication complexes derived from crude extracts of poliovirus-infected cells are active in synthesizing viral RNA in vitro. They contain all three species of viral-specific RNAs (PENMAN et al. 1964; GIRARD et al. 1967; BALTIMORE and GIRARD 1966) and numerous viral-specific proteins, including those responsible for polymerase activity (GIRARD et al. 1967; CALIGUIRI and TAMM 1969, 1970b; CALIGUIRI and MOSSER 1971; MOSSER et al. 1972; CALIGUIRI and COMPANS 1973; BIENZ et al. 1983). Many viral proteins are isolated with the active, crude membrane especially those from the P2 and P3 region of the genome (SEMLER et al. 1982; TAKEGAMI et al. 1983b; TERSHAK 1984) Protein 3D copurifies with the RNA synthesis activity (LUNDQUIST et al. 1974; see below). A close association of polio protein 2C with RNA in replication complexes has been shown both biochemically (RÖDER and KOSCHEL 1975) and by immunocytological electron microscopy techniques (BIENZ et al. 1987). VPg and larger polypeptides which contain VPg sequences are found exclusively with membrane fractions that are active in RNA replication (TAKEGAMI et al. 1983b). The possible involvement of these proteins in viral RNA synthesis will be discussed in a later section.

Purification of active replication complexes from the crude membrane fraction was attempted in order to identify proteins required for RNA synthesis. Solubilization of the polymerase activity was first achieved by use of a deoxycholate NP40 mixed detergent, to produce a 70S complex which was still active in synthesizing polio-specific RNA (EHRENFELD et al. 1970). Subsequent precipitation with 2 M LiCl and centrifugation in sucrose gradients or Sepharose 2B chromatography resulted in additional purification of the polymerase-template complex. RNase A or T1 in high salt did not destroy the template for polymerase activity in these complexes, but RNase III did, implying that at least portions of the template were double-stranded (LUNDQUIST and MAIZEL 1978a). These data are seemingly in conflict with the data of RICHARDS et al. (1984), who analyzed replicative intermediates which were psoralen-cross-linked in vivo and which were shown to be predominantly single stranded. It is likely, however, that the detergent treatment and purification procedures resulted in significant alterations in the structure of the template-enzyme complex. The principal viral protein in the partially purified complex was the polio-encoded protein, $3D^{pol}$ (LUNDQUIST et al. 1974). These complexes were estimated to contain approximately 3000–12000 active polymerase molecules per cell, representing only 1% of the total $3D^{pol}$ molecules produced. LUNDQUIST and MAIZEL (1978a, b) have reported that only the polymerase synthesized early after infection (before $3\frac{1}{4}$ h) becomes associated with the replication complex, and thus only this early form of the enzyme is functional and utilized in poliovirus RNA replication. A mechanism for selection of only a subset of polymerase molecules synthesized at early times of infection for inclusion into replication complexes, however, was not proposed.

3.2 RNA Synthesis by Replication Complexes

Product RNAs synthesized by the membranous complexes in vitro include three separable species—ssRNA, dsRNA, and RI—which resemble the species of viral

RNAs found in vivo. Their kinetics of formation were analyzed by pulse-chase experiments in vitro (GIRARD 1969; McDONNELL and LEVINTOW 1970). RI served as a precursor to ssRNA; dsRNA appeared to be a reaction by-product which accumulated in a linear fashion. RI became labeled rapidly and soon reached steady-state levels, whereas ssRNA accumulated more slowly, eventually reaching a level that exceeded both of the other forms. Detergent treatment of crude replication complexes resulted in a loss of the ability to synthesize ssRNA (ETCHISON and EHRENFELD 1981; CALIGUIRI 1974; McDONNELL and LEVINTOW 1970); after such treatment only RI and dsRNA products were formed. The reason for this change is not understood, but it appears to reflect the extreme importance of the intact architecture of the membrane-bound complex.

3.3 Uridylylation of VPg

All newly synthesized viral RNA recovered from infected cells is covalently linked to the small, virus-encoded protein, VPg; and this linkage is formed either at the time of or immediately following the initiation of new RNA strands (NOMOTO et al. 1977b; PETTERSSON et al. 1978). Crude replication complexes contain an activity that catalyzes the addition of one or two uridylate residues to VPg or a VPg-containing precursor protein, to yield VPgpU and VPgpUpU (TAKEGAMI et al. 1983a). The uridylylation reaction is dependent upon the presence of an endogenous RNA template (TAKEDA et al. 1987). The exact structure of the substrate for the VPg uridylylation reaction has not been defined. TAKEGAMI et al. (1983b) have shown an association of putative precursor forms of VPg, including P2-3AB and 2C-3AB, with membranes in replication complexes. These observations are consistent with a model whereby 2C binds single-stranded segments of replicating RNA and these complexes are anchored in vesicles by hydrophobic regions present in 3A sequences of these same precursors. No uridylylated pre-VPg proteins were detected in these reactions, suggesting that cleavage of pre-VPg occurs concomitantly with the addition of nucleotides to VPg. Under conditions which stimulated the yield of VPgpUpU, it was possible to demonstrate its extension into longer RNA products (TAKEDA et al. 1986). Surprisingly, the reaction conditions found for optimal synthesis of VPgpUpU were different from those for optimal elongation of the VPg-containing RNA strands, suggesting that two distinct processes are involved. Synthesis can be continued on VPgpUpU to produce longer strands, perhaps even genome-length strands, although the stoichiometry of these reactions is difficult to establish.

Pulse-chase experiments have shown that polio RNA strands up to genome length, but not longer, are formed by replication complexes in vitro (TOYODA et al. 1987). Using similar complexes prepared from cells infected with a Sabin strain which is temperature sensitive for viral replication, neither VPgpU nor VPgpUpU accumulated at the restrictive temperature. On the other hand, the formation of ssRNA from RI proceeded with normal kinetics, suggesting a temperature-sensitive step in the initiation of RNA strand synthesis. This effect was attributed to the amino terminal portion of $3D^{pol}$, suggesting that a functional $3D^{pol}$ or 3CD may be needed for uridylylation of VPg and initiation of plus strand RNA synthesis.

4 The Enzyme $3D^{pol}$

Poliovirus-infected cells contain an activity which incorporates ribonucleotides into viral RNA. The activity is insensitive to actinomycin D, is not present in uninfected cells, and it increases with time postinfection in proportion to the increase in appearance of viral proteins. It was therefore assumed that the activity was comprised, at least in part, of a protein encoded by the virus. Initial efforts to identify which viral protein(s) were responsible for the RNA polymerase activity involved the purification of a replication complex from infected cells incubated with [^{35}S]methionine at such times that only viral proteins would become radiolabeled. A partially purified fraction was obtained which demonstrated viral RNA polymerase activity in vitro, and which contained predominantly one viral polypeptide, then designated NCVP 4 or P63, now known as $3D^{pol}$ (LUNDQUIST et al. 1974; FLANEGAN and BALTIMORE 1977). Subsequently the replication complex was demonstrated to contain two viral proteins, designated 4a and 4b ($3D^{pol}$; ETCHISON and EHRENFELD 1980). The protein 4a was unstable and had overlapping tryptic peptides with $3D^{pol}$ (ETCHISON and EHRENFELD 1980), and may be an N-terminally extended precursor to $3D^{pol}$ (SEMLER et al. 1983).

The first tentative identification of a viral protein as an RNA polymerase was subsequently confirmed by an independent approach. Reasoning that the initial step in viral RNA replication was the utilization of the poly(A) sequences at the 3′ end of the virion RNA to form the poly(U) sequence at the 5′ end of negative strand RNA, FLANEGAN and BALTIMORE (1977) sought and found a poly(A)-dependent poly(U) polymerase activity which was solubilized with detergent from the membrane fraction of infected cells. Subsequently, this activity was purified from the soluble fraction of poliovirus-infected HeLa cells, and shown to copurify with viral polypeptide 3D (FLANEGAN and VAN DYKE 1979; DASGUPTA et al. 1979; VAN DYKE and FLANEGAN 1980). This activity coincided with a poliovirus RNA-dependent RNA polymerase activity and thus appeared to be the enzyme responsible for RNA chain elongation during polio RNA replication. Several other independent purification schemes have been used to provide highly purified forms of $3D^{pol}$ (ANDREWS and BALTIMORE 1986; YOUNG et al. 1985; LUBINSKI et al. 1987; PLOTCH et al. 1989).

4.1 Generation of $3D^{pol}$

Complete maturation to polypeptide 3D appears to be required for enzymatic activity. Although uncleaved protein 3CD also copurifies with RNA polymerase activity during initial stages of purification from infected cells, it is resolved from 3D by gel filtration on Sephacryl S-200, and shows no detectable peak of activity (VAN DYKE and FLANEGAN 1980). In addition, 3CD proteins expressed from cloned cDNAs in *E. coli* manifest no polymerase activity unless they are cleaved to 3D (EHRENFELD and RICHARDS 1989). Production of 3D from 3CD or other fusion proteins does occur in *E. coli* as a result of protease activity inherent in the 3C sequence (MORROW et al. 1987; ROTHSTEIN et al. 1988). It is not known, however, whether that cleavage to generate 3D occurs in *cis* via intramolecular proteolysis or in *trans* by an intermolecular reaction. Infected cells accumulate large amounts of 3CD which apparently never gets cleaved. Indeed, it is only assumed that 3CD is a precursor of 3D in vivo. It is possible that 3D is only generated by cleavage of a larger precursor (e.g., 3ABCD), and that once cleavage at the 3B/3C junction occurs, no further cleavage of 3CD can occur. In fact, the mechanism that generates enzymatically active 3D is not known; nor is it known whether the relative levels of 3CD and 3D in the cell are regulated.

4.2 Properties of the Purified Enzyme

Although 3D^{pol} has been purified in several laboratories, both from virus-infected cells and from *E. coli* containing plasmids engineered to express the enzyme, surprisingly little has been learned about its physical structure. It is comprised of a single polypeptide chain of 52 kDa with a pI about 7.4 (VAN DYKE and FLANEGAN 1980) and with an unremarkable predicted amino acid composition. Computer-assisted hydropathicity plots suggest that it is a hydrophilic protein (SKERN et al. 1984), with a relatively high alpha helix content (C. C. BURNS, O. C. RICHARDS and E. EHRENFELD, unpublished observations). A 10 amino acid tract (residues 323 to 332) consisting of a GDD surrounded by hydrophobic residues is conserved among numerous plant and animal virus RNA polymerases and is thought to represent the active site of the polymerase (KAMER and ARGOS 1984). Estimates of the size of the purified, active enzyme by elution from gel filtration columns or by velocity sedimentation in glycerol gradients relative to proteins of known molecular weight are clouded by a high dependence upon ionic strength; nevertheless, it is likely that the active species is a monomer (VAN DYKE and FLANEGAN 1980). Searches for protein modifications, such as glycosylation, acylation, and sulfation, were negative (URZAINQUI and CARRASCO 1989). However, a report by RANSONE and DASGUPTA (1989) stated that a fraction of both 3D^{pol} and 3CD are phosphorylated at serine residues. The implications of the existence of a modified subpopulation of polymerase in infected cells with regard to function may have important bearing on the RNA replicative process. It should be mentioned that the enzyme preparation purified from infected HeLa cells is, by virtue of the purification protocol, precisely that enzyme which is not associated with RNA in the membrane-bound replication complex. Thus, it is not known whether the various properties attributed to the purified enzyme are also exhibited by the enzyme actually engaged in replication.

Catalytic activities associated with 3D^{pol} include an RNA elongation activity, which requires an RNA template and some type of primer. The primer could be replaced by a secondary structure element in the template, e.g., tRNA-like structure, which presumably allows self-priming (TUSCHALL et al. 1982). Elongation of an RNA primer with a free 3′ hydroxyl occurs readily; on the other hand, addition to a tyrosine hydroxyl, e.g., in VPg, by a purified polymerase has neither been demonstrated nor disproven. As long as a suitable oligonucleotide primer is provided, the purified enzyme demonstrates no apparent specificity towards template sequences; therefore, any template specificity inherent in the replication reaction must reside in additional proteins or activities which mediate the initiation step of the reaction. The elongation rate is highly pH dependent, varying from 83 nuc/min at pH 7.0, 30 °C, 3 mM Mg^{2+} to 635 nuc/min at pH 8.0, 30 °C, 7 mM Mg^{2+} (VAN DYKE et al. 1982). Other laboratories, utilizing different enzymes preparations, reported 375 nuc/min at pH 8.0, 30 °C, 5 mM Mg^{2+} (LUBINSKI et al. 1987) or 500 nuc/min at pH 8.0, 30 °C, 3.5 mM Mg^{2+} (BARON and BALTIMORE 1982d). The error frequency of the polymerization reaction is 10^{-3}–10^{-4} (WARD et al. 1988); this frequency is affected by the Mg^{2+} concentration, shifting upward fivefold upon increasing from 3 mM to 7 mM Mg^{2+}. In addition to catalyzing phosphodiester bond formation directed by an RNA template, 3D^{pol} is likely to manifest other activities. These may include nucleotide binding, RNA binding, interaction with other host or viral proteins or membrane components, and contribution to the protease activity of 3CD (YPMA-WONG et al. 1988; JORE et al. 1988). Some preliminary binding studies with poliovirion RNA have been performed, but the responsible sites on the template and/or enzyme have not been elucidated, and the specificity of the binding is unclear (OBERSTE and FLANEGAN 1988). A single report of an additional activity associated with 3D^{pol} expressed in E. coli described an apparent terminal transferase activity which facilitates formation of product strands up to twice genome length (PLOTCH et al. 1989); the possible significance of this activity in polio RNA replication is discussed in a subsequent section. Finally, some role for 3D^{pol} in the uridylylation of VPg or VPg precursors has been suggested, but the precise activity contributed by 3D^{pol} in this reaction has not been defined (TOYODA et al. 1987).

4.3 3D^{pol} Produced from Cloned cDNA

Three laboratories have produced catalytically active 3D^{pol} from cloned cDNA in E. coli (MORROW et al. 1987; RICHARDS et al. 1987; ROTHSTEIN et al. 1988; PLOTCH et al. 1989). In one case, the enzyme was designed to contain a single additional amino acid, the N-terminal methionine (PLOTCH et al. 1989). MORROW et al. (1987) cloned a more extensive sequence, 3A*BCD (where the asterisk denotes an incomplete protein sequence), as a fusion protein which underwent self-cleavage to release active polymerase. RICHARDS et al. (1987) expressed active enzyme from a variety of constructs which generated mature 3D^{pol} as a result of proteolytic processing by 3C^{pro} either in *cis* or in *trans*. The availability of these clones might enable investigators to identify functional domains of the enzyme by site-directed mutagenesis of the cDNA sequence. To date, single amino acid substitutions in or very near to the putative active site have generated enzyme with reduced or no RNA polymerase activity (C. MORROW, personal communication). A spontaneous deletion of a trinucleotide that codes for a tryptophan residue normally present as the fifth amino acid from the N-terminus abolished all detectable polymerase activity, as did deletion of the C-terminal 53 amino acids of the normal protein (PLOTCH et al. 1989). A more extensive analysis of a series of proteins containing single amino acid insertions at various positions scattered throughout the length of the protein also generally yielded inactive enzyme (BURNS et al. 1989). One mutant, however, with a leucine insertion after amino acid residue 257 showed about 10% of wild-type polymerase activity. Further study of this mutant may reveal some structure-function relationships. In this study, the bacterial extracts were also examined for 3CD protease activity, using viral capsid protein precursor as substrate. Some of the mutations in the 3D portion of 3CD resulted in loss of protease activity. The number and locations of the amino acid alterations did not permit the identification of a separate domain that contributes to the protease activity in the 3D protein, however. Continued systematic analyses of mutations introduced into the 3D sequence may eventually reveal functions of this protein other than RNA chain elongation. In addition, 3D^{pol} has recently been produced in cultured insect cells from a recombinant baculovirus (K. NEUFELD, O.C. RICHARDS, and E. EHRENFELD, unpublished). It is hoped that the yields and properties of this expression system will permit a more extensive biochemical analysis than has been possible previously.

4.4 Lessons from Genetics

Since the construction of infectious cDNA clones of the poliovirus genome, a number of defined mutants have been genetically engineered. A very large proportion of these mutants are nonviable, and thus relatively uninformative, although it is likely that eventually this route will prove fruitful. At present, very few viable viruses have been produced by deliberate mutagenesis of the 3D^{pol} coding sequence; none have been well characterized.

An approach classically used to direct or to supplement biochemical studies of enzymatic functions is the isolation and genetic analysis of viable or conditional-lethal mutant viruses resulting from chemical or radiation-induced mutagenesis, or isolated as spontaneously recurring variants. In the case of polio or other picornaviruses, genetic analysis is complicated by the fact that the genome represents a single cistron, and mutations in one region of the genome may affect numerous protein products derived from other coding regions, resulting in highly pleiotropic effects. For example, mutations which alter protease function may affect several or all of the proteins which are products of protease cleavage. Other types of mutations may produce alterations in polyprotein structure which consequently may be processed aberrantly. Multiple mutations in a single genome frequently accumulate, and identifying and mapping those responsible for a given phenotype is extremely laborious. In addition, very few genetic markers exist for identifying phenotypes. Some temperature-sensitive (ts) virus mutants that fail to synthesize RNA at restrictive temperatures have been isolated, but the precise location or number of the mutations has not been identified (COOPER 1977; HEWLETT et al. 1982).

A classical, long-term genetic study has been in progress for the last decade by GIRARD and coworkers. A number of ts mutants of poliovirus type 1 were isolated after mutagenesis with 5-fluorouracil or nitrous acid. Two ts viruses with defective RNA replication were isolated which proved to result from mutations in the 3D coding sequence. One of these, ts 035, replaced asparagine 424 with aspartate (AGUT et al. 1989). Viral protein synthesis appeared normal at the nonpermissive temperature, and no defect in RNA elongation was detected when the activity of crude replication complexes or of purified polymerase was measured in vitro at 39°C. The authors suggested that the amino acid change in $3D^{pol}$ impaired the initiation step of viral RNA synthesis, but no further activities (such as uridylylation of VPg) were tested, and the putative defective initiation reaction has not been identified. The defect is not likely to be caused by an effect of the nucleotide substitution on the RNA template structure, since the mutant can be complemented by either wild-type or another ts which can supply functional viral protein in *trans*.

A second independent mutation in $3D^{pol}$ was isolated which manifested a temperature-dependent inhibition of viral RNA synthesis by actinomycin D (KEAN et al. 1989). This phenotype was due to a glutamine to histidine substitution at residue 170 of $3D^{pol}$. Poliovirus strains sensitive to actinomycin D have been observed to result from mutations at other locations on the genome, including the 5' noncoding region (RACANIELLO and MERIAM 1986; TRONO et al. 1988). This indicates that the actinomycin D phenotype can be caused by alterations in the RNA structure. The actinomycin-sensitivity of the $3D^{pol}$ mutant, however, is probably not due to RNA structure alterations, since the inhibition by actinomycin D was increased when cells were pretreated with the drug before infection. Interestingly, even wild-type poliovirus replication was partially inhibited after preincubation of cells for 3 h with actinomycin D. This suggests that the drug does not act directly on viral replication complexes, but rather that it may act by reducing the level(s) of some cellular factor(s) required for viral RNA replication. Apparently, the mutation in the 3D polymerase protein may decrease the affinity of the enzyme for the (undefined) host factor(s), effectively increasing the factor requirement in a manner which is accentuated by elevated temperature. These data provide some of the strongest evidence for the involvement of host factor(s) in poliovirus RNA replication.

One additional $3D^{pol}$ mutant was engineered by site-directed mutagenesis to yield a single amino acid insertion after residue 354 in the 3D code (BERNSTEIN et al. 1986). This virus produced small plaques at all temperatures, and RNA synthesis was delayed, although normal levels of RNA were eventually produced. As with mutations in several other nonstructural protein coding regions, this mutant was unable to be complemented by 3D functions provided in *trans* by viruses that contained mutations in other genes. Again, the authors felt that it was not likely that the nucleotide insertions disrupted RNA structure so as to impair its ability to serve as template, although such an event would explain the *cis*-acting nature of the mutation. Instead, they postulated that RNA replication occurs in a complex of proteins which remain associated with the RNA from which they were translated; thus, the functions of these proteins (3D, 2B, perhaps 2C) could not be supplied in *trans*. As mentioned above, however, the ts 035 mutant was shown to be efficiently rescued in complementation tests with either wild-type virus or another ts mutant (AGUT et al. 1989). At least some of the conflicting data regarding *cis* and *trans* functions might be the result of different mechanisms of generation of plus and minus strand replicative intermediates. For example, a 3D molecule might be restricted to transcribe the plus strand template from which it was translated; the subsequent attachment to a minus strand template, however, could occur by diffusion and de novo binding. Clearly, it will be necessary to study additional mutants in order to understand the failure to demonstrate complementation of one, but not another, 3D mutant.

5 Other Viral Proteins Involved in RNA Synthesis

Studies described in the previous section confirmed that viral protein 3D^{pol} was the major polio protein involved in RNA chain elongation. The complete RNA replication reaction, however, involves additional steps of initiation and VPg addition, regulation of synthesis of minus versus plus strands, interaction of the replication complex with intracellular membranes, etc. A number of viral proteins are specifically associated with membranes (TERSHAK 1984), and many of these have been implicated as being involved in the process of RNA replication.

5.1 3CD and 3C

Some data have been interpreted to indicate the involvement of 3CD in the polio RNA replication reaction, despite the absence of elongation activity associated with 3CD (VAN DYKE and FLANEGAN 1980). These studies showed, that stabilization of 3CD by the addition of protease inhibitor to infected cells correlated with a lengthened period of RNA synthesis compared with cells not treated with the inhibitor (KORANT 1975); in addition, inhibition of RNA synthesis by a type 3 strain of poliovirus at restrictive temperature correlated with a breakdown of 3CD (BOWLES and TERSHAK 1978).

Protease 3C catalyzes most of the proteolytic cleavages in the viral polyprotein to generate functional proteins. It is therefore not surprising that 3C sequences affect viral RNA replication due to the need to generate 3D^{pol}. For example, chemical mutagenesis has been used to generate temperature-sensitive mutants in polio type 1 RNA (BELLOCQ et al. 1984). One particularly interesting mutant, ts 247, fails to make RNA at restrictive temperatures. By fingerprint analysis this mutant has one change in 3C and one in 3D. When the 3C mutation was inserted into an infectious cDNA and transfected into cells, viral RNA synthesis was significantly reduced (KEAN et al. 1988). Normal 3B/3C cleavage occurred, but 3C/3D cleavage was only 20% of wild-type levels. A similar mutant was described by DEWALT and SEMLER (1987). Thus, under some conditions, RNA replication may be regulated by the cleavage of 3CD to 3D. Interestingly, 2B/2C cleavage was also greatly reduced in ts 247. If 2C is important in RNA chain initiation (see below), reduced 2C generation may also be responsible for reduced levels of RNA synthesis.

5.2 3B (VPg) and 3AB

The genome-linked protein, VPg (3B), is somehow involved in viral RNA replication, since it is attached to the 5′ ends of nascent RNA chains. This observation suggested that VPg may be a primer for initiation of RNA synthesis; alternatively it may be attached at the time of or shortly after the initiation step. As discussed above, a uridylylated form of VPg has been shown to be produced in vitro by crude complexes which synthesized viral RNA, and this same nucleotidyl-VPg moiety has been observed in infected cells (CRAWFORD and BALTIMORE 1983), although its role in the initiation of viral RNA synthesis has not been elucidated.

Several polypeptides containing VPg sequences have been detected in extracts of infected cells by immunoprecipitation with antibodies directed against synthetic peptides corresponding to regions of VPg (BARON and BALTIMORE 1982a; SEMLER et al. 1982; TAKEGAMI et al. 1983b). These antibodies immunoprecipitated a 12–14 kDa protein which had VPg sequences at its carboxy terminus (presumably 3AB) as well as a VPg immunoreactive species of 45–48 kDa (presumably 2C3AB), and an even larger species (presumably P2-3AB). These proteins are considered to be precursors of VPg. They are enriched in membrane fractions that include active replication complexes (SEMLER et al. 1982; TAKEGAMI et al. 1983b), and they include a hydrophobic domain present in the 3A sequences that might serve as the membrane-anchoring region. An additional viral polypeptide, most likely corresponding to 2C3ABC, has also been identified by KEAN et al. (1988). It is not known whether one or more of these precursor proteins serves as the direct donor in the VPg addition reaction, or whether protein cleavage occurs first and free VPg is directly linked to the RNA. Interestingly, two independent studies reported that some form of VPg precursor became associated with product RNA synthesized by 3D^{pol} from a polio RNA template in an in vitro reaction in the presence of one or more host proteins (BARON and BALTIMORE 1982b; MORROW et al. 1984). Product RNA which was digested with RNase or alkali yielded residual material which was immunoprecipitated with anti-VPg serum and had mobilities on SDS-polyacrylamide gels consistent with proteins of 47–49 kDa and 14 kDa (MORROW et al. 1984). Whatever the nature of the VPg donor in these studies, it must have been present as a contaminant in the preparation of 3D^{pol} used for RNA polymerization.

A likely candidate for the donor of VPg to the membrane-associated viral RNA replication complex is the smallest VPg-containing polypeptide, 3AB. It is a relatively stable and abundant product in infected cells (SEMLER et al. 1982; TAKEGAMI et al. 1983b). A cold-sensitive mutant constructed to contain a single inserted serine residue at amino acid position 15 of 3A caused reduced synthesis of both plus and minus strands at 32° C (BERNSTEIN et al. 1986; BERNSTEIN and BALTIMORE 1988). Another mutant, in which the wild-type amino acid threonine 67 of 3A was changed to isoleucine, manifested temperature-sensitive growth in HeLa cells, which appeared to result from a primary lesion in RNA synthesis other than chain elongation (GIACHETTI et al. 1989). Selective mutagenesis of the VPg sequence itself had variable effects on virus viability (KUHN et al. 1988), presumably caused by effects on RNA replication and/or protein processing.

5.3 2C

Other VPg-containing polypeptides (2C3AB, 2C3ABC, or P23AB) may function as VPg precursors and may be involved in the formation of the membrane-bound replication complex. Protein 2C, by itself, is a very abundant and stable viral product. It has been reported to be present in a partially purified polio RNA replication complex (RÖDER and KOSCHEL 1975). More importantly, 2C has been demonstrated to be the viral gene product responsible for virus sensitivity to guanidine. Millimolar concentrations of guanidine inhibit polio RNA synthesis, very likely at an initiation step (CALIGUIRI and TAMM 1973; BALTIMORE 1968; TERSHAK 1982). The inhibition results in loss of production of genome-length ssRNA, dsRNA, and replicative intermediates. Variants of poliovirus are guanidine resistant, and efforts were made to map this trait on the poliovirus genome by genetic recombination (TOLSKAYA et al. 1983; EMINI et al. 1984) and by examination of electrophoretic variants in guanidine-resistant mutants (ANDERSON-SILLMAN et al. 1984). These studies established that guanidine resistance mapped to the nonstructural protein coding region of polio RNA. It remained for PINCUS et al. (1986) to definitively pinpoint the guanidine resistance trait to a specific site in 2C. In fact, guanidine-dependent mutants contain two mutations in 2C, one at the precise site mentioned above for guanidine resistance and a second mutation at an independent site. Recently LI and BALTIMORE (1988) examined polioviruses which resulted from transfection of HeLa cells with plasmids containing linker-insertion mutations at defined sites in 2C. These mutations led to greatly reduced titers of temperature-sensitive virus which was defective for RNA synthesis at restrictive temperatures. In shift-up experiments RNA synthesis was rapidly shut down, suggesting that 2C is needed continually for RNA synthesis. These mutations could be complemented in *trans* by coinfection with viruses carrying mutations in other genes. The precise role protein 2C plays in polio RNA replication has not been determined.

5.4 *Cis* vs *Trans* Activities

Several investigators have observed the generation of defective-interfering (DI) particles in preparations of poliovirus by multiple serial passages at high multiplicities of infection (COLE et al. 1971; LUNDQUIST et al. 1979; NOMOTO et al. 1979; KAJIGAYA et al. 1985). These particles all contain viral RNAs with deletions in the capsid protein coding region. Purified DI particles can infect and initiate a replication cycle but they are unable to synthesize capsid proteins; therefore, their replication requires coinfection with helper virus which provides capsid proteins in *trans* for DI genome packaging and assembly. KUGE et al. (1986) performed detailed structural analyses of cloned cDNAs of DI genomes. They observed that the deletions in every DI genome retained the correct reading frame for viral protein synthesis. These results suggested that one or all of the viral nonstructural proteins was required to act in *cis* at some stage in viral replication. This was consistent with the results of genetic complementation studies which showed that a mutation in 3D^{pol} was not rescued by coinfection with a virus containing a wild-type 3D gene, and thus at least some 3D function could not be supplied in *trans* (BERNSTEIN et al. 1986). More recently, KAPLAN and RACANIELLO (1988) constructed DNAs with in-frame deletions in the capsid region in infectious cDNA. Transcripts of these DNAs (called "replicons") were transfected into HeLa cells and this RNA was replicated, viral proteins were synthesized, and normal processing occurred. HAGINO-YAMAGISHI and NOMOTO (1990) have developed this system to produce DI particles by superinfection of the transfected cells with wild-type virus. In this way they could demonstrate conclusively that replication of the replicon required in-frame deletions, implying that at least one *cis*-acting viral protein must play a role in the DI RNA replication and this function could not be supplied by the superinfecting virus. (The formal possibility exists that the apparent requirement for *cis*-acting nonstructural protein in RNA replication rather reflects a requirement for concurrent translation of the RNA in order to present a proper substrate for initiation of RNA synthesis.) Extension of this approach, with genetically manipulated replicons, should prove to be a powerful tool for deducing the specific role(s) of required viral proteins.

6 Viral RNA Template Recognition

Replication of the parental virion RNA involves synthesis of a negative strand and then production of new plus strands. The replication machinery, therefore, must be capable of initiating synthesis at two very different termini: the 3' end of the plus strand template is a rather long stretch of adenylate residues, whereas the 3' end of the minus strand template is a heteropolymeric sequence terminating most often in two A residues (LARSEN et al. 1980; RICHARDS and EHRENFELD 1980). It is generally assumed that the ends of the template RNAs contain primary and/or secondary structures that serve as recognition signals for the polymerase, or for a VPg donor serving as primer, or for a host factor which may help catalyze the initiation reaction. No specific signals have yet been identified.

The production of infectious polio RNA transcripts from cloned cDNA was reported several years ago (VAN DER WERF et al. 1986). The initial construct generated a polio plus strand transcript with 60 extra nucleotides to the 5' side of the viral RNA sequences and 626 nucleotides of pBR322 sequence beyond an 84-nucleotide poly(A) tract at the 3' end. This transcript had a specific infectivity about 0.1% of RNA isolated from virus. Subsequently, a T7 promoter was placed so that viral RNA was produced containing only two additional guanylate residues at the 5' end and only seven extra nucleotides past the poly(A) tract at the 3' end. The specific infectivity of this transcript was increased to 5% of virion RNA. More recently, SARNOW (1989) constructed templates for in vitro transcription by T7 RNA polymerase which contained the same extra two guanylate residues at the 5' end, but which terminated in a long (100 nucleotide) homopolymeric poly(A) tract. These RNAs were equally as infectious as virion RNA. Short poly(A) tails (12 nucleotides) or long heteropolymeric sequences at the 3' end of the transcripts reduced the infectivity of the RNA molecules. Earlier studies had shown that viral RNAs with 3' terminal poly(A) sequences shorter than 20 nucleotides had much reduced infectivity (SPECTOR and BALTIMORE 1974). Although none of these studies indicated whether the reduced infectivity of the terminally modified viral RNAs was due to RNA instability or to their being unfavorable templates for translation or for initiation of RNA synthesis, they did emphasize the importance of both 5' and 3' termini in at least one step in virus replication.

More information has become available from specific site-directed mutagenesis of the 3' and 5' noncoding regions of viral RNA. An 8-nucleotide insertion in the 3' noncoding region of poliovirus type 1 generated virus that was temperature-sensitive for RNA synthesis (SARNOW et al. 1986). As expected, this mutation was cis-acting, and could not be rescued by coinfection with virus with unaltered 3' noncoding sequences (BERNSTEIN et al. 1986). The 3' noncoding regions of all three serotypes of poliovirus are highly conserved. Similarly, mutations in the 5' noncoding region of viral RNA, which would be reflected in the 3' end of the negative strand template, have also been shown to affect viral RNA synthesis (SEMLER et al. 1986; RACANIELLO and MERIAM 1986; DILDINE and

SEMLER 1989). The continued analysis of this type of mutation, as well as of replicons (transcripts of engineered viral cDNAs that have been deleted for sequences not required for their own replication), should prove useful in the dissection and identification of noncoding sequence signals that are important for replication protein recognition or binding. These replicons have been shown to be very potent *trans* inhibitors of viral RNA replication, presumably by serving as competitors for the binding of these proteins (KAPLAN and RACANIELLO 1988).

7 Reconstitution of RNA Synthesis In Vitro

The initial step toward efforts to reconstitute the poliovirus RNA replication process in vitro entailed purification of the viral polymerase ($3D^{pol}$) from the soluble portion of the cytoplasm from poliovirus-infected HeLa cells. As the enzyme was more extensively purified active polymerization was shown to require an RNA template, NTPs, and a primer, in addition to $3D^{pol}$. That is, purified $3D^{pol}$ had only an elongation activity; it could not initiate RNA strand synthesis.

7.1 Host Factor Activities

A major contribution in purification of the polio polymerase ($3D^{pol}$) was the use of oligo(U) as a synthetic primer on the poliovirion RNA template for assay of enzyme activity (FLANEGAN and VAN DYKE 1979). This convenient assay, however, gave no indication of the nature of the natural primer for poliovirus RNA synthesis. Since the purified enzyme was unable to initiate by itself, a search was made for some other proteins that would allow $3D^{pol}$ to initiate synthesis of a minus strand. DASGUPTA et al. (1980) first reported a host protein in the ribosomal salt wash of uninfected HeLa cells that could restore polymerase activity to purified polio polymerase using a polio RNA template in the absence of oligo(U). The host factor (HF) was purified (BARON and BALTIMORE. 1982c; DASGUPTA 1983a) as a 67 000–69 000 kDa protein. HF showed an avid association with $3D^{pol}$ on polymerase-Sepharose affinity columns, and antibodies to HF both inhibited the polymerization reaction in vitro, and coprecipitated $3D^{pol}$ (DASGUPTA 1983b). These results suggest a tight association between the two proteins, prior to purification. Product RNA, synthesized in the presence of HF, appeared to be full-length, complementary RNA strands. Some of these molecules precipitated with anti-VPg antibodies and thus appeared to be directly linked to VPg-related proteins, which were apparently present in the $3D^{pol}$ preparations (BARON and BALTIMORE 1982b; MORROW et al. 1984).

MORROW et al. (1985) reported that purification of HF coincided with a phosphorylated protein and a protein kinase activity. Anti-HF simultaneously inhibited the kinase activity and the HF-polymerase RNA synthesis activity. This kinase activity could phosphorylate the translational initiation factor, eIF-2; however, the biological significance of this reaction is unknown. This HF preparation, in conjunction with the viral polymerase, was capable of producing infectious plus strand RNA from a negative strand template that was synthesized by SP6 RNA polymerase from cloned cDNA (KAPLAN et al. 1985). Neither the nature of the initiation reaction nor the precise termini of the product plus strands were determined.

In another laboratory, utilizing different preparations of polymerase and HF, YOUNG et al. (1985) found that HF could support synthesis of complementary RNA on a poliovirus RNA template, with formation of a heterogeneously sized population of RNA strands, a small portion of which were twice the length of genome RNA. The authors suggested that HF-polymerase activity stimulated formation of snapback structures resulting from covalent attachment of product to template RNA (YOUNG et al. 1986). Anti-VPg antibody immunoprecipitated product RNA only by virtue of its attachment to template which contained VPg, and if VPg was removed from the template before transcription, no product was immunoprecipitated with anti-VPg antibody (YOUNG et al. 1986, 1987). The polymerase preparation used in these studies was likely a more highly purified preparation than that used by MORROW et al. (1984), where VPg-related proteins must have been present as an inadvertent contaminant.

A possible mechanism for the generation of snapback structures was provided by ANDREWS et al. (1985), who utilized a 68 kDa HF preparation that had terminal uridylyl transferase (TUT) activity. They proposed that template and product snapback molecules could result from the terminal addition of uridylate residues to the 3' poly(A) tract of the template, followed by formation of a 3' terminal hairpin which would create a self-priming sequence for chain elongation by $3D^{Pol}$. It was claimed that about four UMP residues were added to the polio RNA template in the absence of $3D^{Pol}$, presumably at the 3' terminus (ANDREWS and BALTIMORE 1986), although quantitation was not provided. The sizes or properties of the products of transcription of viral RNA by TUT and $3D^{Pol}$ were not presented. Interestingly, a recent study of purified viral polymerase expressed in *E. coli* suggests that TUT activity may be associated with the polymerase itself (PLOTCH et al. 1989).

An independent study of the products of a HF-polymerase reaction confirmed the synthesis of a heterogeneously sized population of product strands, up to approximately twice the length of the template (HEY et al. 1986, 1987). These products, however, appeared to be generated by an endonuclease activity present in the HF preparation which seemingly nicks the template RNA at random positions, generating 3' termini that can prime complementary strand synthesis from the nick sites. Their conclusions were supported by biochemical data which demonstrated that the poly(A) tract was removed from the template by HF, and that oxidation of the 3' terminal nucleotide of the template did not prevent its utilization, as would be expected for a TUT-catalyzed hairpin formation. LUBINSKI et al. (1986, 1987) have also described the formation in vitro of snapback structures generated by endonucleolytic nicking. Products of both plus and minus strand polarities were detected in these studies, suggesting that under in vitro conditions, product RNAs may again snapback on themselves to prime synthesis of the complementary strand. No detectable TUT activity was observed in the HF preparations used in these studies.

A small portion of virus-specific double-stranded RNA that is found in infected cells also appears to be snapback hairpins, up to dimer length molecules (YOUNG et al. 1985; RICHARDS et al. 1987a). However, one study showed that these hairpins also lacked poly(A) sequences, and thus were likely formed after nucleolytic removal of the 3' end of plus-strand templates (RICHARDS et al. 1987a).

The results of the in vitro reactions reconstituted from purified polymerase and host factor preparations have been disappointing. The products may be genome-length or heterogeneous and up to dimer length; synthesis of strands of a single polarity or of both polarities have been reported; host factors of some sort are clearly required to initiate a reaction in vitro, but their activities vary, and their biochemical roles have not been well defined. Many of the contrasting observations are likely due to impure preparations of $3D^{pol}$ and/or HF, leading to extraneous activities, and part of the difficulty is likely due to the absence of additional proteins which may be required for the reaction to proceed.

Although it is generally believed that the tissue tropism and host-cell specificity of poliovirus replication is largely determined by the distribution of cell-specific receptors for virus adsorption and entry, it is possible that the availability of host factors required for RNA replication may affect the efficiency of virus growth in different cells. It has been shown that virus replication in established human blood cell lines varies with the cell lineage and step of differentiation (OKADA et al. 1987; ROIVAINEN and HOVI 1989), although RNA synthesis is probably not the variable step in these cells. Such variations could, of course, result from cellular differences in factors required for translation, assembly, or other processes, as well as for RNA synthesis. It remains for the roles of specific HF activities to be identified and analyzed in different cells to show whether cell-specific RNA replication reactions might be important in the pathogenesis of human infections.

7.2 VPg Addition

None of the data generated by studies of RNA synthesis in vitro have explained the mechanism by which VPg is linked to the nascent RNA. In the hairpin model put forth by FLANEGAN and coworkers, VPg addition to the newly synthesized strand must occur subsequent to or concomitant with an endonucleolytic cleavage at the hairpin. A preliminary report by TOBIN et al. (1989) described the successful linkage of synthetic VPg to poliovirus minus strand RNA synthesized in vitro from virion RNA template, by purified $3D^{pol}$ and HF, in a reaction previously shown to produce snapback RNAs up to dimer size in length. The

VPg-linked product was heterogeneous in size, up to the size of virion RNA. The linkage of VPg to product RNA required no other protein, and no energy source. The authors proposed that covalent linkage of VPg to the 5′ end of the newly synthesized minus strand occurs by a transesterification mechanism that simultaneously cleaves at the template-product junction and forms a covalent bond between the tyrosine hydroxyl of VPg and the phosphate of the 5′ terminal UMP residue. It was further suggested that this reaction may be catalyzed by a sequence within the template RNA itself. If it is confirmed that this reaction can occur in vitro, with specificity for polio RNA, it will be extremely exciting to evaluate its role during viral RNA replication in vivo, and to determine whether VPg itself serves as the substrate for this reaction, or a VPg-containing precursor protein. In another laboratory, HF-mediated reactions that generated genome-length, complementary RNAs in vitro appeared to have been initiated with a large VPg precursor of 47–49 kDa (MORROW et al. 1984). Since these molecules were only fortuitously present as contaminants in the polymerase preparation, the mechanism of their linkage could not be studied.

A fundamentally different model for VPg addition is based on the utilization of VPg or a precursor to VPg, most likely in a uridylylated form, as the direct primer for initiation of RNA synthesis (LEE et al. 1977; FLANEGAN et al. 1977; BARON and BALTIMORE 1982b; MORROW et al. 1984; WIMMER 1982; CRAWFORD and BALTIMORE 1983; TOYODA et al. 1987; TAKEDA et al. 1987). Support for this model is based on studies that demonstrate that VPgpUpU is present in infected cells (CRAWFORD and BALTIMORE 1983) and that it is synthesized in vitro by a crude, membranous replication complex (TAKEGAMI et al. 1983a). Furthermore, at least some VPgpUpU in the crude replication complex can be extended into VPg-RNA strands in vitro (TAKEDA et al. 1986). The enzyme(s) for uridylylation of VPg has not been identified, and the mechanism of utilization of the nucleotidyl protein as primer is unknown. Of special interest is whether there is specificity for polio RNA in this proposed priming reaction.

7.3 Models of RNA Chain Initiation

Figure 1 illustrates two models for viral RNA synthesis in the membranous replication complex in infected cells. Panel a shows a VPg precursor, embedded in the membrane via the hydrophobic region of 3A, and uridylylated to generate a nucleotidyl protein, which serves as a primer for the synthesis of product RNA. As synthesis is initiated, 3C or 3CD cleaves the 3A/3B junction, leaving VPg (3B) on the 5′ end of the nascent strand, and 3D then continues the elongation reaction. VPg uridylylation could be catalyzed by a host enzyme, or possibly by 3CD or 3D, itself. Product RNA is released from the template either by the inherent topography of the membrane-bound components, or by a ssRNA binding protein or helicase activity. The function of 2C in this or any other specific step in the reaction has not been defined; and specific requirements for host factors, if any, remain to be elucidated.

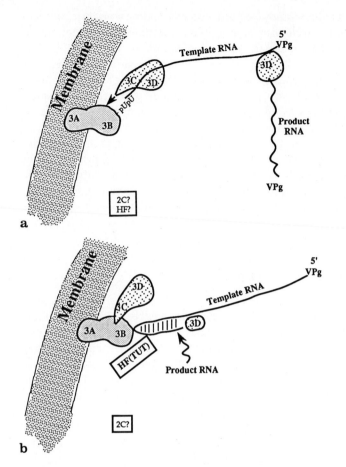

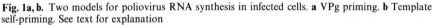

b

Fig. 1a, b. Two models for poliovirus RNA synthesis in infected cells. **a** VPg priming. **b** Template self-priming. See text for explanation

An illustration of a template RNA self-priming mechanism is shown in panel b. 3AB is again embedded in the membrane. A host factor, suggested to be TUT, generates and perhaps stabilizes the formation of a hairpin at the 3′ end of the template, which is elongated by 3D. 3AB is cleaved by 3C or 3CD, and simultaneously the hairpin is cut and VPg (3B) is linked to the newly formed 5′ end. The role of 2C is still undefined. This model would likely generate a dsRNA product, which then must be unwound for further replication.

7.4 Synthesis of Plus and Minus Strands

To date, no in vitro system has been assembled which can utilize a plus strand template and produce new plus strand progeny. Indeed, it is not known whether synthesis of minus and plus strands from their corresponding templates proceeds by the same or different mechanism(s) and utilizes the same or different factors and components. Studies of crude replication complexes generally involve synthesis of new plus strands, since this is the major endogenous reaction occurring in infected cells. On the other hand, the majority of the attempts at reconstituted reactions start with plus strand template, since this is the most readily available. Thus, the various models commonly found in the literature are not mutually exclusive; synthesis of minus strands could occur by hairpin priming from a HF-induced structural modification of plus strand templates, while synthesis of plus strands could occur by direct nucleotidyl-VPg priming in the membrane-associated replication complex. Similarly numerous other proteins (viral and cellular) might be required to effect a complete reaction in vitro, while the efficiencies and side products of partial reactions might continue to prove quite difficult to interpret.

8 Summary and Future Directions

Progress in the field of poliovirus RNA replication has been slow and tedious, extending over the past 25 years. A picture has emerged of a highly complex structure of viral transcription units composed of a viral RNA template strand with multiple nascent product strands, the viral protein 3D catalyzing the elongation of these strands, and numerous other viral and very likely cellular proteins, involved in the initiation of new strand synthesis. The entire transcription unit is anchored in an extensive, newly assembled membrane structure, with which many of the essential viral proteins are very tightly associated. The issue of how VPg becomes linked to the 5' ends of nascent RNA strands is still unresolved, as is the biochemical role of viral protein 2C. Clearly, the identification of any host factors that may be required for RNA replication is of paramount importance, along with an understanding of their precise role.

The crude membranous replication complexes isolated from infected cells were an attractive starting point for biochemical analyses, because they synthesize in vitro all of the viral RNA species found in vivo; and they therefore are presumed to contain all the components necessary for the complete reaction. The major reaction occurring in these complexes is synthesis of new plus strands from a minus strand template. Indeed, the site of synthesis of minus strands has not been determined, and it is not known whether the reaction mechanism and the proteins required for synthesis of the two strands are the same. Aside from detecting the uridylylation of a VPg-containing protein, however, the crude complexes have proved difficult to work with and they have not yielded partial reaction information. The integrity of the membrane appears essential for biochemical activity.

Studies of the enzyme $3D^{pol}$ are likely to progress rapidly in the near future, since the cDNA has been cloned and expressed in several systems, yielding enzymatically active protein. Production of large quantities of the protein should now be possible, so that purification and subsequent biochemical analysis of functional domains, RNA contact sites, and amino acid residues comprising the active site(s) can be examined. These studies will be greatly aided if the three-dimensional structure of the polymerase can be solved.

The ability to produce viral genomes with known mutations in specific locations by genetic manipulation of cloned cDNA has already increased our knowledge of viral gene products important for RNA replication. This technology also provides a tool to examine the template signals utilized for recognition and binding by the polymerase and/or accessory proteins.

Perhaps most uncertain is the immediate prognosis for studies of in vitro reactions reconstituted from purified components. Although partial reactions can be observed with a partial contingent of components, individual reaction rates and products can become quite distorted by the absence of the other players needed to complete the pathway. Purification can eliminate essential activities or include contaminating ones, either of which may affect the products formed. Some investigators in the field are convinced that the hydrophobic environment provided by cellular membranes, where polypeptides can be topologically anchored and processed, is an essential factor in the replication process. It is of particular importance to collect evidence that the reactions studied in vitro are a true reflection of what occurs in vivo during viral RNA synthesis. Without such evidence, the studies become an exercise in enzyme characterization, determining what kinds of reactions are capable of occurring, under the conditions of the in vitro assays. It has already become clear, however, that application of the knowledge gained by a combination of all of the genetic and biochemical approaches available will ultimately be required to provide new insights into how poliovirus replicates its RNA.

References

Agut H, Kean KM, Fichot O, Morasco J, Flanegan JB, Girard M (1989) A point mutation in the poliovirus polymerase gene determines a complementable temperature-sensitive defect of RNA replication. Virology 168: 302–311

Alexander HE, Koch G, Mountain IM, Van Damme O (1958) Infectivity of ribonucleic acid from poliovirus in human cell monolayers. J Exp Med 108: 493–506

Ambros V, Baltimore D (1978) Protein is linked to the 5′ end of poliovirus RNA by phosphodiester linkage to tyrosine. J Biol Chem 253: 5263–5266

Anderson-Sillman K, Bartal S, Tershak DR (1984) Guanidine-resistant poliovirus mutants produce modified 37-kilodalton proteins. J Virol 50: 922–928

Andrews NC, Baltimore D (1986) Purification of a terminal uridylyl transferase that acts as host factor in the in vitro poliovirus replicase reaction. Proc Natl Acad Sci USA 83: 221–225

Andrews NC, Levin D, Baltimore D (1985) Poliovirus replicase stimulation by terminal uridylyl transferase. J Biol Chem 260: 7628–7635

Baltimore D (1964) In vitro synthesis of viral RNA by the poliovirus RNA polymerase. Proc Natl Acad Sci USA 51: 450–456

Baltimore D (1966) Purification and properties of poliovirus double-stranded ribonucleic acid. J Mol
    Biol 18: 421–428
Baltimore D (1968) Inhibition of poliovirus replication by guanidine. In: Sanders M, Lennette EH
    (eds) Medical and applied virology .Green, St Louis, pp 340–347
Baltimore D, Girard M (1966) An intermediate in the synthesis of poliovirus RNA. Proc Natl Acad Sci
    USA 56: 741–748
Baltimore D, Franklin RM, Eggers HJ, Tamm I (1963) Poliovirus induced RNA polymerase and the
    effects of virus-specific inhibitors on its production. Proc Natl Acad Sci USA 49: 843–849
Baltimore D, Girard M, Darnell JE (1966) Aspects of the synthesis of poliovirus RNA and the
    formation of virus particles. Virology 29: 179–189
Baron MH, Baltimore D (1982a) Antibodies against the chemically synthesized genome-linked
    protein of poliovirus react with native virus-specific proteins. Cell 28: 395–404
Baron MH, Baltimore D (1982b) Anti-VPg antibody inhibition of the poliovirus replication reaction
    and production of covalent complexes of VPg-related proteins and RNA. Cell 30: 745–752
Baron MH, Baltimore D (1982c) Purification and properties of a host cell protein required for
    poliovirus replication in vitro. J Biol Chem 257: 12351–12358
Baron MH, Baltimore D (1982d) In vitro copying of viral positive strand RNA by poliovirus replicase.
    Characterization of the reaction and its products. J Biol Chem 257: 12359–12366
Bellocq C, Agut H, Van Der Werf S, Girard M (1984) Biochemical characterization of poliovirus type
    1 temperature-sensitive mutants. Virology 139: 403–407
Bernstein HD, Baltimore D (1988) Poliovirus mutant that contains a cold-sensitive defect in viral
    RNA synthesis. J Virol 62: 2922–2928
Bernstein HD, Sarnow P, Baltimore D (1986) Genetic complementation among poliovirus mutants
    derived from an infectious cDNA clone. J Virol 60: 1040–1049
Bienz K, Egger D, Rasser Y, Bossart W (1983) Intracellular distribution of poliovirus proteins and the
    induction of virus-specific cytoplasmic structures. Virology 131: 39–48
Bienz K, Egger D, Pasamontes L (1987) Association of polioviral proteins of the P2 genomic region
    with the viral replication complex and virus-induced membrane synthesis as visualized by electron
    microscopic immunochemistry and autoradiography. Virology 160: 220–226
Bishop JM, Koch G (1967) Purification and characterization of poliovirus induced infectious double-
    stranded RNA. J Biol Chem 242: 1736–1743
Bishop JM, Koch G (1969) Infectious replicative intermediate of poliovirus: purification and
    characterization. Virology 37: 521–534
Bishop JM, Koch G, Evans B, Merriman M (1969) Poliovirus replicative intermediate: structural
    basis of infectivity. J Mol Biol 46: 235–249
Bowles SA, Tershak DR (1978) Proteolysis of non-capsid protein 2 of type 3 poliovirus at the
    restrictive temperature: breakdown of non-capsid protein 2 correlates with loss of RNA synthesis.
    J Virol 27: 443–448
Burns CC, Lawson MA, Semler BL, Ehrenfeld E (1989) Effects of mutations in poliovirus 3D^{pol} on
    RNA polymerase activity and on polyprotein cleavage. J Virol 63: 4866–4874
Caliguiri LA (1974) Analysis of RNA associated with the poliovirus RNA replication complexes.
    Virology 58: 526–535
Caliguiri LA, Compans RW (1973) The formation of poliovirus particles in association with the RNA
    replication complexes. J Gen Virol 21: 99–108
Caliguiri LA, Mosser AG (1971) Proteins associated with the poliovirus RNA replication complex.
    Virology 46: 375–386
Caliguiri LA, Tamm I (1969) Membranous structures associated with translation and transcription
    of poliovirus RNA. Science 166: 885–886
Caliguiri LA, Tamm I (1970a) The role of cytoplasmic membranes in poliovirus biosynthesis.
    Virology 42: 100–110
Caliguiri LA, Tamm I (1970b) Characterization of poliovirus-specific structures associated with
    cytoplasmic membranes. Virology 42: 112–122
Caliguiri LA, Tamm I (1973) Guanidine and 2-(α-hydroxybenzyl) benzimidazole (HBB): selective
    inhibitors of picornavirus multiplication. In: Carter W (ed) Selective inhibitors of viral function.
    CRC, Cleveland, pp 257–294
Cole CN, Smoler D, Wimmer E, Baltimore D (1971) Defective interfering particles of poliovirus I.
    Isolation and physical properties. J Virol 7: 478–485
Cooper PD (1977) Genetics of picornaviruses. In: Fraenkel-Conrat H, Wagner RR (eds) Comprehens-
    ive virology, vol 9. Plenum, New York, pp 133–207

Crawford NM, Baltimore D (1983) Genome-linked protein VPg of poliovirus is present as free VPg and VPgpUpU in poliovirus-infected cells. Proc Natl Acad Sci USA 80: 7452–7455

Crocker TT, Pfendt E, Spendlove R (1964) Poliovirus: growth in non-nucleate cytoplasm. Science 145: 401–403

Darnell JE, Levintow L, Thoren MM, Hooper JL (1961) The time course of synthesis of poliovirus RNA. Virology 13: 271–279

Darnell JE, Girard M, Baltimore D, Summers DF, Maizel JV (1967) The synthesis and translation of poliovirus RNA. In: Colter JS, Paranchych W (eds) The molecular biology of viruses. Academic, New York, pp 375–401

Dasgupta A (1983a) Purification of host factor required for in vitro transcription of poliovirus RNA. Virology 127: 245–251

Dasgupta A (1983b) Antibody to host factor precipitates poliovirus RNA polymerase from poliovirus-infected HeLa cells. Virology 128: 252–259

Dasgupta A, Baron MH, Baltimore D (1979) Poliovirus replicase: a soluble enzyme able to initiate copying of poliovirus RNA. Proc Natl Acad Sci USA 76: 2679–2683

Dasgupta A, Zabel P, Baltimore D (1980) Dependence of the activity of the poliovirus replicase on a host cell protein. Cell 19: 423–429

Dewalt PG, Semler BL (1987) Site-directed mutagenesis of proteinase 3C results in a poliovirus deficient in synthesis of viral RNA polymerase. J Virol 61: 2162–2170

Dildine SL, Semler BL (1989) The deletion of 41 proximal nucleotides reverts a poliovirus mutant containing a temperature-sensitive region in the 5' noncoding region of genomic RNA. J Virol 63: 847–862

Ehrenfeld E, Richards OC (1989) Studies of poliovirus RNA polymerase expressed in E. coli: attempts to understand virus RNA replication. In: Semler BL, Ehrenfeld E (eds) Molecular aspects of picornavirus infection and detection. Amer. Society of Microbiology, Washington DC, pp 95–105

Ehrenfeld E, Maizel JV, Summers DF (1970) Soluble RNA polymerase complex from poliovirus-infected HeLa cells. Virology 40: 840–846

Emini EA, Leibowitz J, Diamond DC, Bonin J, Wimmer E (1984) Recombinants of Mahoney and Sabin strain poliovirus type 1: analysis of in vitro phenotypic markers and evidence that resistance to guanidine maps in the non-structural proteins. Virology 137: 74–85

Etchison D, Ehrenfeld E (1980) Viral polypeptides associated with the RNA replication complex in poliovirus-infected cells. Virology 107: 135–142

Etchison D, Ehrenfeld E (1981) Comparison of replication complexes synthesizing poliovirus RNA. Virology 111: 33–46

Fernandez-Munoz R, Darnell JE (1976) Structural difference between the 5' termini of viral and cellular mRNA in the poliovirus-infected cells: possible basis for the inhibition of host protein synthesis. J Virol 18: 719–726

Flanegan JB, Baltimore D (1977) Poliovirus-specific primer-dependent RNA polymerase able to copy poly(A). Proc Natl Acad Sci USA 74: 3677–3680

Flanegan JB, Baltimore D (1979) Poliovirus polyuridylic acid polymerase and RNA replicase have the same viral polypeptide. J Virol 29: 352–360

Flanegan JB, Van Dyke TA (1979) Isolation of a soluble and template-dependent poliovirus RNA polymerase that copies virion RNA in vitro. J Virol 32: 155–161

Flanegan JB, Pettersson RF, Ambros V, Hewlett MJ, Baltimore D (1977) Covalent linkage of a protein to a defined nucleotide sequence at the 5' terminus of virion and replicative intermediate RNAs of poliovirus. Proc Natl Acad Sci USA 74: 961–965

Franklin RM, Baltimore D (1962) Patterns of macromolecular synthesis in normal and virus-infected mammalian cells. Cold Spring Harb Symp Quant Biol 27: 175–198

Giachetti C, Semler BL (1989) In: 2nd International symposium on positive strand RNA viruses, Vienna.

Girard M (1969) In vitro synthesis of poliovirus ribonucleic acid: role of the replicative intermediate. J Virol 3: 376–384

Girard M, Baltimore D, Darnell JE (1967) The poliovirus replication complex: site for synthesis of poliovirus RNA. J Mol Biol 24: 59–74

Hagino-Yamagishi K, Nomoto A (1990) In vitro construction of poliovirus defective-interfering particles. J Virol 63: 5386–5392

Hewlett MJ, Rose JK, Baltimore D (1976) 5'-Terminal structure of poliovirus polyribosomal RNA is pUp. Proc Natl Acad Sci USA 73: 327–330

Hewlett MJ, Axelrod JH, Antinoro N, Feld R (1982) Isolation and preliminary characterization of temperature-sensitive mutants of poliovirus type 1. J Virol 41: 1089–1094

Hey TD, Richards OC, Ehrenfeld E (1986) Synthesis of plus- and minus-strand RNA from poliovirion RNA template in vitro. J Virol 58: 790–796

Hey TD, Richards OC, Ehrenfeld E (1987) Host factor-induced template modification during synthesis of poliovirus RNA in vitro. J Virol 61: 802–811

Holland JJ, McLaren LC, Syverton JT (1959) The mammalian cell-virus relationship. IV. Infection of naturally insusceptible cells with enterovirus ribonucleic acid. J Exp Med 110: 65–80

Jore J, Gens BD, Jackson RD, Pouwels PH, Enger-Valk B (1988) Poliovirus protein 3CD is the active protease for processing of the precursor protein P1 in vitro. J Gen Virol 69: 1627–1636

Kajigaya S, Arakawa H, Kuge S, Koi T, Imura N, Nomoto A (1985) Isolation and characterization of defective-interfering particles of poliovirus Sabin I strain. Virology 142: 307–316

Kamer G, Argos P (1984) Primary structural comparison of RNA-dependent polymerases from plant, animal and bacterial viruses. Nucleic Acids Res 12: 7269–7282

Kaplan G, Racaniello VR (1988) Construction and characterization of poliovirus subgenomic replicons. J Virol 62: 1687–1696

Kaplan G, Lubinski J, Dasgupta A, Racaniello VR (1985) In vitro synthesis of infectious poliovirus RNA. Proc Natl Acad Sci USA 82: 8424–8428

Kean KM, Agut H, Fichot O, Wimmer E, Girard M (1988) A poliovirus mutant defective for self-cleavage at the COOH-terminus of the 3C protease exhibits secondary processing defects. Virology 163: 330–340

Kean KM, Agut H, Fichot O, Girard M (1989) Substitution in the poliovirus replicase genes determines actinomycin D sensitivity of viral replication at elevated temperature. Virus Res 12: 19–32

Kitamura N, Semler BL, Rothberg PG, Larsen GR, Adler CJ, Dorner AJ, Emini EA, Hanecak R, Lee JJ, Van Der Werf S, Anderson CW, Wimmer E (1981) Primary structure, gene organization, polypeptide expression of poliovirus RNA. Nature 291: 547–553

Korant BD (1975) Regulation of animal virus replication by protein cleavage. In: Reich E, Rifkin BD, Shaw E (eds) Proteases and biological control. Cold Spring Harbor Laboratory, Cold Spring Harbor pp 621–635

Kuge S, Saito L, Nomoto A (1986) Primary structure of poliovirus defective interfering particle genomes and possible generation mechanism of the particles. J Mol Biol 192: 473–487

Kuhn RJ, Tada H, Ypma-Wong MF, Semler BL, Wimmer E (1988) Mutational analysis of the genome-linked protein VPg of poliovirus. J Virol 62: 4207–4215

La Monica N, Meriam C, Racaniello VR (1986) Mapping of sequences required for mouse neurovirulence of poliovirus type 2 Lansing. J Virol 57: 515–525

Larsen GR, Dorner AJ, Harris TJR, Wimmer E (1980) The structure of poliovirus replicative form. Nucleic Acids Res 8: 1217–1229

Lee YF, Nomoto A, Detjen BM, Wimmer E (1977) The genome-linked protein of picornaviruses 1. A protein covalently linked to poliovirus genome RNA. Proc Natl Acad Sci USA 74: 59–63

Li J-P, Baltimore D (1988) Isolation of poliovirus 2C mutants defective in viral RNA synthesis. J Virol 62: 4016–4021

Lubinski JM, Kaplan G, Racaniello VR, Dasgupta A (1986) Mechanism of in vitro synthesis of covalently linked dimeric RNA molecules by the poliovirus replicase. J Virol 58: 459–467

Lubinski JM, Ransone LJ, Dasgupta A (1987) Primer-dependent synthesis of covalently linked dimeric RNA molecules by poliovirus replicase. J Virol 61: 2997–3003

Lundquist RE, Maizel JV (1978a) Structural studies of the RNA component of the poliovirus replication complex I. Purification and biochemical characterization. Virology 85: 434–444

Lundquist RE, Maizel JV (1978b) In vivo regulation of the poliovirus RNA polymerase. Virology 89: 484–493

Lundquist RE, Ehrenfeld E, Maizel JV (1974) Isolation of a viral polypeptide associated with poliovirus RNA polymerase. Proc Natl Acad Sci USA 71: 4773–4777

Lundquist RE, Sullivan M, Maizel JV (1979) Characterization of a new isolate of poliovirus defective interfering particles. Cell 18: 759–769

McDonnell JP, Levintow L (1970) Kinetics of appearance of the products of poliovirus-induced RNA polymerase. Virology 42: 999–1006

Meyer J, Lundquist RE, Maizel JV (1978) Structural studies of the RNA component of the poliovirus replication complex. Virology 85: 445–455

Montagnier L, Sanders FK (1963) Sedimentation properties of infective ribonucleic acid extracted from encephalomyocarditis virus. Nature 197: 664–669

Morrow CD, Navab M, Peterson C, Hocko J, Dasgupta A (1984) Antibody to poliovirus genome-linked protein (VPg) precipitates in vitro synthesized RNA attached to VPg-precursor polypeptide(s). Virus Res 1: 89–100

Morrow CD, Gibbons GF, Dasgupta A (1985) The host protein required for in vitro replication of poliovirus is a protein kinase that phosphorylates eukaryotic initiation factor-2. Cell 40: 913–921

Morrow CD, Warren B, Lentz MR (1987) Expression of enzymatically active poliovirus RNA-dependent RNA polymerase in Escherichia coli. Proc Natl Acad Sci USA 84: 6050–6054

Mosser AG, Caliguiri LA, Scheid AS (1972) Chemical and enzymatic characteristics of cytoplasmic membranes of poliovirus-infected HeLa cells. Virology 47: 30–38

Noble J, Levintow L (1970) Dynamics of poliovirus-specific RNA synthesis and the effects of inhibitors of virus replication. Virology 40: 634–642

Nomoto A, Lee YF, Wimmer E (1976) The 5'-end of poliovirus mRNA is not capped with m^{7}G(5') ppp(5')Np. Proc Natl Acad Sci USA 73: 375–380

Nomoto A, Kitamura N, Golini F, Wimmer E (1977a) The 5'-terminal structures of poliovirion RNA and poliovirus mRNA differ only in the genome-linked protein VPg. Proc Natl Acad Sci USA 74: 5345–5349

Nomoto A, Detjen B, Pozzatti R, Wimmer E (1977b) The location of the polio genome protein in viral RNAs and its implication for RNA synthesis. Nature 268: 208–213

Nomoto A, Jacobsen A, Lee YF, Dunn J, Wimmer E (1979) Defective interfering particles of poliovirus: mapping of deletion and evidence that the deletion in genome of D1 (1), (2) and (3) are located in the same region. J Mol Biol 128: 179–196

Nomoto A, Omata T, Toyoda H, Kuge S, Horie H, Kataoka Y, Genba Y, Nakano Y, Imura N (1982) Complete nucleotide sequence of the attenuated poliovirus Sabin 1 strain genome. Proc Natl Acad Sci USA 79: 5793–5797

Oberste MS, Flanegan JB (1988) Measurement of poliovirus RNA polymerase binding to poliovirion and nonviral RNAs using a filter-binding assay. Nucleic Acids Res 16: 10339–10352

Okada Y, Toda G, Oka H, Nomoto A, Yoshikura H (1987) Poliovirus infection of established human blood cell lines. Relationship between the differentiation stage and susceptibility or cell killing. Virology 156: 238–245

Penman S, Becker Y, Darnell JE (1964) A cytoplasmic structure involved in the synthesis and assembly of poliovirus components. J Mol Biol 8: 541–555

Pettersson RF, Ambros V, Baltimore D (1978) Identification of a protein linked to nascent poliovirus RNA and to the polyuridylic acid of negative-strand RNA. J Virol 27: 357–365

Pincus SE, Diamond DC, Emini EA, Wimmer E (1986) Guanidine-selected mutants of poliovirus: mapping of point mutations of polypeptide 2C. J Virol 57: 638–646

Plotch SJ, Palant O, Gluzman Y (1989) Purification and properties of poliovirus RNA polymerase expressed in Escherichia coli. J Virol 63: 216–225

Racaniello VR, Baltimore D (1981) Molecular cloning of poliovirus cDNA and determination of the complete nucleotide sequence of the viral genome. Proc Natl Acad Sci USA 78: 4887–4891

Racaniello V, Meriam C (1986) Poliovirus temperature-sensitive mutant containing a single nucleotide deletion in the 5' non-coding region of the viral RNA. Virology 155: 498–507

Ransone LJ, Dasgupta A (1989) Multiple isoelectric forms of poliovirus RNA-dependent RNA polymerase: evidence for phosphorylation. J Virol 63: 4563–4568

Richards OC, Ehrenfeld E (1980) Heterogeneity of the 3' end of minus strand RNA in poliovirus replicative form. J Virol 36: 387–394

Richards OC, Ehrenfeld E, Manning J (1979) Strand-specific attachment of avidin spheres to double-stranded poliovirus RNA. Proc Natl Acad Sci USA 76: 676–680

Richards OC, Martin SC, Jense HG, Ehrenfeld E (1984) Structure of poliovirus replicative intermediate RNA. Electron microscope analysis of RNA cross-linked in vivo with psoralen derivative. J Mol Biol 173: 325–340

Richards OC, Hey TD, Ehrenfeld E (1987a) Poliovirus snapback double-stranded RNA isolated from infected HeLa cells is deficient in poly(A). J Virol 61: 2307–2310

Richards OC, Ivanoff LA, Bienkowska-Szewczyk K, Butt B, Petteway SR, Rothstein MA, Ehrenfeld E (1987b) Formation of poliovirus RNA polymerase 3D in Escherichia coli by cleavage of fusion proteins expressed from cloned viral cDNA. Virology 161: 348–356

Röder A, Koschel A (1975) Virus-specific proteins associated with the replication complex of poliovirus RNA. J Gen Virol 14: 846–852

Roivainen M, Hovi T (1989) Replication of poliovirus in human mononuclear phagocyte cell lines is dependent on the stage of cell differentiation. J Med Virol 27: 91–94

Rothberg PG, Harris TJR, Nomoto A, Wimmer E (1978) The genome-linked protein of picornaviruses. V.O^4-(5'-Uridylyl)-tyrosine is the bond between the genome-linked protein and the RNA of poliovirus. Proc Natl Acad Sci USA 75: 4868–4872

Rothstein MA, Richards OC, Amin C, Ehrenfeld E (1988) Enzymatic activity of poliovirus RNA polymerase synthesized in *Escherichia coli* from viral cDNA. Virology 164: 301–308

Roy P, Bishop DHL (1970) Isolation and properties of poliovirus minus strand-ribonucleic acid. J Virol 6: 604–609

Rueckert RR, Wimmer E (1984) Systematic nomenclature of picornavirus proteins. J Virol 50: 957–959

Sarnow P (1989) Role of 3'-end sequences in infectivity of poliovirus transcripts made in vitro. J Virol 63: 467–470

Sarnow P, Bernstein HS, Baltimore D (1986) A poliovirus temperature-sensitive mutant located in a non-coding region of the genome. Proc Natl Acad Sci USA 83: 571–575

Savage T, Granboulan N, Girard M (1971) Architecture of the poliovirus replicative intermediate RNA Biochimie 53: 533–543

Scharff MD, Thoren MM, McElvain NF, Levintow L (1963) Interruption of poliovirus RNA synthesis by p-fluorophenylalanine and puromycin. Biochem Biophys Res Commun 10: 127–132

Semler BL, Anderson CW, Hanecak R, Dorner LF, Wimmer E (1982) A membrane-associated precursor to poliovirus VPg identified by immunoprecipitation with antibodies directed against a synthetic heptapeptide. Cell 28: 405–412

Semler BL, Hanecak R, Dorner LF, Anderson CW, Wimmer E (1983) Poliovirus RNA synthesis in vitro: Structural elements and antibody inhibition. Virology 126: 624–633

Semler BL, Johnson VH, Tracy S (1986) A chimeric plasmid from cDNA clones of poliovirus and coxsackievirus produces a recombinant virus that is temperature-sensitive. Proc Natl Acad Sci USA 83: 1777–1781

Skern T, Sommergruber W, Blaas D, Pieler C, Kuechler E (1984) Relationship of human rhinovirus strain 2 and poliovirus as indicated by comparison of the polymerase gene regions. Virology 136: 125–132

Spector DH, Baltimore D (1974) Requirement of 3'-terminal poly(adenylic acid) for the infectivity of poliovirus RNA. Proc Natl Acad Sci USA 83: 2330–2334

Spector DH, Baltimore D (1975a) Polyadenylic acid on poliovirus RNA. II. Poly(A) on intracellular RNAs. J Virol 15: 1418–1431

Spector DH, Baltimore D (1975b) Polyadenylic acid on poliovirus RNA. IV. Poly(U) in replicative intermediate and double-stranded RNA. Virology 67: 498–505

Stanway G, Cann AJ, Hauptmann R, Hughes P, Clarke LD, Mountford RC, Minor PD, Schild GC, Almond JW (1983) The nucleotide sequence of poliovirus type 3 leon 12 a, b: comparison with poliovirus type 1. Nucleic Acids Res 11: 5629–5643

Takeda N, Kuhn RJ, Yang CF, Takegami T, Wimmer E (1986) Initiation of poliovirus plus-strand RNA synthesis in a membrane complex of infected HeLa cells. J Virol 60: 43–53

Takeda N, Yang CF, Kuhn RJ, Wimmer E (1987) Uridylylation of the genome-linked protein of poliovirus in vitro is dependent upon an endogenous RNA template. Virus Res 8: 193–204

Takegami T, Kuhn RJ, Anderson CW, Wimmer E (1983a) Membrane-dependent uridylylation of the genome-linked protein VPg of poliovirus. Proc Natl Acad Sci USA 80: 7447–7451

Takegami T, Semler BL, Anderson CW, Wimmer E (1983b) Membrane fractions active in poliovirus RNA replication contain VPg precursor polypeptides. Virology 128: 33–47

Tershak DR (1982) Inhibition of poliovirus polymerase by guanidine in vitro. J Virol 41: 313–318

Tershak DR (1984) Association of poliovirus proteins with the endoplasmic reticulum. J Virol 52: 777–783

Tobin GJ, Young DC, Flanegan JB (1989) Self-catalized linkage of poliovirus terminal protein VPg to poliovirus RNA. Cell 59: 511–519

Tolskaya EA, Romanova LA, Kolesnikova MS, Agol VI (1983) Intertypic recombination in poliovirus: genetic and biochemical studies. Virology 124: 121–132

Toyoda H, Kohara M, Kataoka Y, Suganuma T, Omata T, Imura N, Nomoto A (1984) Complete nucleotide sequences of all three poliovirus serotype genomes. Implication for genetic relationship, gene function and antigenic determinants. J Mol Biol 174: 561–585

Toyoda H, Yang C-F, Takeda N, Nomoto A, Wimmer E (1987) Analysis of RNA synthesis of type 1 poliovirus by using an in vitro molecular genetic approach. J Virol 61: 2816–2822

Trono D, Andino R, Baltimore D (1988) An RNA sequence of hundreds of nucleotides at the 5' end of poliovirus RNA is involved in allowing viral protein synthesis. J Virol 62: 2291–2299

Tuschall DM, Hiebert E, Flanegan JB (1982) Poliovirus RNA-dependent RNA polymerase synthesizes full-length copies of poliovirion RNA, cellular mRNA, and several plant virus RNAs in vitro. J Virol 44: 209–216

Urzainqui A, Carrasco L (1989) Post-translational modifications of poliovirus proteins. Biochem Biophys Res Commun 158: 263–271

Van der Werf S, Bradley J, Wimmer E, Studier FW, Dunn J (1986) Synthesis of infectious poliovirus RNA by purified T7 RNA polymerase. Proc Natl Acad Sci USA 83: 2330–2334

Van Dyke TA, Flanegan JB (1980) Identification of poliovirus polypeptide p63 as a soluble RNA-dependent RNA polymerase. J Virol 35: 732–740

Van Dyke TA, Rickles RJ, Flanegan JB (1982) Genome-length copies of poliovirion RNA are synthesized in vitro by the poliovirus RNA dependent RNA polymerase. J Biol Chem 257: 4610–4617

Ward CD, Stokes MAM, Flanegan JB (1988) Direct measurement of the poliovirus RNA polymerase error frequency in vitro. J Virol 62: 558–562

Wimmer E (1982) Genome-linked proteins of viruses. Cell 28: 199–201

Wu M, Davidson N, Wimmer E (1978) An electron microscope study of proteins attached to poliovirus RNA and its replicative form (RF). Nucleic Acids Res 5: 4711–4723

Yogo Y, Wimmer E (1972) Polyadenylic acid at the 3'-terminus of poliovirus RNA. Proc Natl Acad Sci USA 69: 1877–1882

Yogo Y, Wimmer E (1973) Poly(A) and poly(U) in poliovirus double-stranded RNA. Nature New Biol 242: 171–174

Yogo Y, Wimmer E (1975) Sequence studies of poliovirus RNA III. Polyuridylic acid and polyadenylic acid as components of the purified poliovirus replicative intermediate. J Mol Biol 92: 467–477

Young DC, Tuschall DM, Flanegan JB (1985) Poliovirus RNA-dependent RNA polymerase and host cell protein synthesize product RNA twice the size of poliovirus RNA in vitro. J Virol 54: 256–264

Young DC, Dunn BM, Tobin GJ, Flanegan JB (1986). Anti-VPg antibody precipitation of product RNA synthesized in vitro by the poliovirus polymerase and host factor is mediated by VPg on the poliovirion RNA template. J Virol 58: 715–723

Young DC, Tobin GJ, Flanegan JB (1987) Characterization of product RNAs synthesized in vitro by poliovirus RNA polymerase purified by chromatography on hydroxylapatite or poly(U) sepharose. J Virol 61: 611–614

Ypma-Wong MF, Dewalt PG, Johnson VH, Lamb JG, Semler BL (1988) Protein 3CD is the major poliovirus proteinase responsible for cleavage of the P1 capsid precursor. Virology 166: 265–270

Zimmerman EF, Heeter M, Darnell JE (1963) RNA synthesis in poliovirus-infected cells. Virology 19: 400–408

Antigenic Structure of Picornaviruses

P. D. Minor

National Institute for Biological Standards and Control, Blanche Lane, South Mimms, Potters Bar,
Herts EN6 3QG, England

Current Topics in Microbiology and Immunology, Vol. 161
© Springer Verlag Berlin Heidelberg 1990

1 Introduction

The detailed antigenic structure of the picornaviruses has been intensively studied over the past few years using a variety of different methods which have frequently given conflicting results. The resolution of the atomic structures of at least one member of each of the four genera of picornavirus has contributed greatly to the accepted views.

The four genera which make up the picornavirus family are the enteroviruses, typified by poliovirus types 1, 2 and 3 and coxsackie B virus; the rhinoviruses, such as rhinovirus 14 and rhinovirus 2; the aphthoviruses or foot and mouth disease viruses, and the cardioviruses, such as mengovirus or encephalomyocarditis virus. Two other viruses will be discussed, namely hepatitis A virus, which has been classified as an enterovirus, and Theiler's virus of mice. Hepatitis A virus has unusual properties for an enterovirus, being an extremely hardy virus which probably does not grow to any significant degree in the human gut, and the sequence of its genomic RNA has little or no homology with those of other enteroviruses studied so far. Theiler's virus similarly has a sequence which suggests that it is more closely related to the cardioviruses than to the enteroviruses, although like the enteroviruses the genome lacks a polycytidylate tract.

Electron microscopy has shown that the virions of all picornaviruses are approximately 27 nm in diameter, not penetrated by the strains used and generally featureless, although a number of techniques have demonstrated their icosahedral symmetry. The positive sense RNA genome of 7.4 to 8.5 kilobases (kb) is enclosed in a protein shell made up of 60 copies each of four proteins, designated VP1, VP2, VP3 and VP4, α, β, γ and δ or, in the formal nomenclature, ID, IB, IC and IA (RUECKERT and WIMMER 1985). The protein shell does not include a lipid membrane although molecules of lipid form an integral part of the structure. It is known that protein VP4 (IA) is myristylated in all picornaviruses examined to date (CHOW et al. 1987) and there is evidence for another lipid species in poliovirus particles at an abundance of one molecule per protomer or 60 copies per virion (FILMAN et al. 1989). The absence of a lipid membrane contributes to the rigidity of the virion and is presumably a factor in the success achieved in crystallising picornaviruses for X-ray crystallographic study. The rigidity of the structure also has implications for the grosser antigenic properties of the viruses and their subunits.

2 General Antigenic Properties of Picornaviruses

Picornaviruses of a particular genus can usually be divided into antigenically distinct serotypes. There are 68 recognised types of enterovirus, namely the 3 serotypes of poliovirus, 23 serotypes of coxsackie A virus (1–22 and 24), 6 serotypes of coxsackie B virus (1–6), 31 types of echovirus, and 5 enteroviruses

(68–72) of which enterovirus 72 is the hepatitis A virus. While the classification is ultimately based on antigenic properties there are equally major differences in host range and pathology, as well as the cellular receptor sites recognised. In contrast there are 100 serotypes of rhinovirus with very similar host range and pathology which may be classified into two groups depending on the receptor site they utilise. Foot and mouth disease viruses are classified into seven serotypes: A, O, C, SAT_1, SAT_2, SAT_3 and Asia 1, and the cardioviruses are generally considered to be monotypic.

As for most if not all virus species studied to date, it is possible to demonstrate antigenic variation between strains of a picornavirus serotype using monoclonal antibodies. The real significance of intratypic strain variations is not clear; all isolates of poliovirus are classified by definition into one of the three existing serotypes and while some may be atypical (MAGRATH et al. 1986) the existing vaccine strains which are over 30 years old function effectively where they are used correctly. This implies that antigenic variation within a serotype is of minimal epidemiological significance. Strain variation within an aphthovirus serotype on the other hand is regarded as a major factor in the production of effective vaccines (RWEYEMAMU and HINGLEY 1984).

The antigenic structure of the picornavirus particle depends on the particle examined and how it has been treated, and has been reviewed elsewhere (MINOR 1987). Relatively slight abuses such as heating at 56 °C can produce major conformational changes in the virion (LE BOUVIER 1955; HUMMELER and TUMILOWICZ 1960; MAYER et al. 1957; ROMBAUT et al. 1983). The data are consistent with the view that the infectious particle is a metastable structure which may be transformed into a more stable noninfectious form with or without loss of the internal protein VP4 (JOKLIK and DARNELL 1961; LONBERG HOLM et al. 1975; BREINDL 1971; LONBERG HOLM and BUTTERWORTH 1976; LONBERG HOLM and YIN 1973; ROMBAUT et al. 1983). The phenomenon has been reported for poliovirus (MAYER et al. 1957), rhinovirus (LONBERG HOLM and YIN 1973), coxsackie B virus (FROMHAGEN 1965) and foot and mouth disease virus (ROWLANDS et al. 1975).

The isolated proteins of picornaviruses induce a poor neutralising antibody response compared with the intact virus. However, isolated VP1 of FMDV has been consistently reported to induce neutralising antibody (MELOEN et al. 1979) and a genetically engineered construct has been shown to induce a protective immune response (KLEID et al. 1981). The data reported for poliovirus are less consistent, some reporting that no proteins are effective, others that VP1 or all proteins may induce a neutralising response (MELOEN et al. 1979; CHOW and BALTIMORE 1982; BLONDEL et al. 1982; DERNICK et al. 1983; VAN DER MAREL et al. 1983). In one paper (EMINI et al. 1983a) the internal protein VP4 was reported to induce neutralising antibody which reacted only with VP3. Capsid protein VP2 of coxsackie B virus has been reported to induce high levels of neutralising antibody (BEATRICE et al. 1980).

Many of the antigenic sites on poliovirus have proved to have strong conformational elements and the somewhat inconclusive data produced with isolated proteins are therefore not surprising. Even where an isolated protein has been shown to be effective, however, as in the case of foot and mouth disease virus, other methods have been required to identify the precise sites of the virus to which antibodies bind. These include:

1. Analysis of the sequence of structural proteins using methods such as that of HOPP and WOODS (1981) to identify hydrophilic nonconserved sequences which are likely to be surface features varying under immune pressure
2. The use of synthetic peptides of a sequence derived from likely antigenic sites to induce an immune response
3. The reaction of monoclonal antibodies with synthetic or genetically engineered peptides of a sequence derived from likely antigenic sites
4. The isolation of mutant viruses escaping neutralisation by monoclonal antibodies
5. X-ray crystallographic resolution of the atomic structure of the virus to identify prominent features likely to be accessible to antibodies

A range of structural features have been identified as antigenic sites on picornaviruses, and the nature of the site has determined the success achieved with the different approaches. It is likely that comparable sites exist on other antigens so that the results obtained with picornaviruses are likely to have a general significance for the study of the antigenic structure of other virus species or antigens.

3 Poliovirus

3.1 Structure

The virion of the type 1 poliovirus strain Mahoney consists of 60 copies each of three main capsid proteins VP1, VP2, and VP3 arranged with icosahedral symmetry, and 60 copies of the small internal protein VP4 which can be considered as a terminal extension of VP2 (HOGLE et al. 1985). The cores of the three large structural proteins are essentially identical, being composed of eight antiparallel strands, forming a wedge-shaped barrel-like structure termed the beta barrel, whose structure is virtually identical for each of the three major proteins although they differ significantly in amino acid sequence. The thin end of the VP1 wedge is directed towards the pentameric apex of the icosahedral particle, while the equivalent portions of VP2 and VP3 alternate about the pseudo-sixfold axis of symmetry at the centre of the triangular face of the icosahedron. The precise orientation of the beta barrels varies but the overall plan of the core structure is the same for all picornaviruses studied to date and essentially the same as for the small spherical plant viruses (HARRISON et al. 1978).

While the cores are structurally very similar, the regions of sequence which link the strands comprising the beta barrel and the N and C terminal sequences vary from protein to protein, and this plays a major part in giving the proteins and the viruses their unique structural features. For polio these linking sequences form several distinct prominent features, one around the pentameric apex composed of four separate loops from VP1 separated by a marked dip or canyon from a complex around the pseudo-sixfold axis, made up of various loops from VP2, VP1 and VP3. The capsid proteins intertwine extensively such that portions of the sequence which are either on different capsid proteins or well separated within a protein may be found together in the final structure. Similarly regions which are close together in the sequence may be oriented in such a way as to make them contribute to different structures in the virion.

The obvious surface features to which ligands such as antibodies are likely to have access include some extended largely free-standing loops, but also some short sequences which interact in complex ways; even the supposedly free-standing loops appear to interact with other features to a greater or lesser extent, as will be discussed below in the context of antigenic chimaeras. The antigenic structure of poliovirus deduced by the isolation of antigenic variants is more consistent with the atomic structure than that obtained by the use of short peptides. It seems reasonable to suppose that this is due in large part to the fixed complex conformational nature of most of the accessible surface features, which may not be readily imitated by short synthetic amino acid sequences.

3.2 Genetic Approaches to the Determination of the Antigenic Structure of Poliovirus

Mutants which are resistant to monoclonal antibodies which neutralise the parental virus have been extensively used to characterise the antigenic sites on the polioviruses. If a mutant is resistant to two separate antibodies it is argued that the mutated amino acid forms part of the epitopes recognised by both antibodies; the epitopes therefore overlap and are said to form part of the same antigenic site. Conversely if all of the mutants resistant to one antibody are still sensitive to a second, and vice versa, the two antibodies recognise completely independent nonoverlapping antigenic sites which do not share amino acids. The mutations may be identified by limited sequencing of the genomic RNA of mutants through likely regions identified by other means, or by complete sequencing of the region encoding the capsid region.

This strategy for analysing the antigenic structure of a virus is only directly applicable if the antibodies neutralise the parent virus and represent all specificities of interest, and if the mutations occur at the site to which the antibody binds without greatly affecting the viability of the virus. These assumptions are not universally valid but the results suggest that they are reasonable working hypotheses.

Extensive studies were performed on type 3 poliovirus using mutants isolated with a large panel of monoclonal antibodies generated from both rats and mice

(FERGUSON et al. 1984; MINOR et al. 1983, 1985; EVANS et al. 1983) and it was possible to show that 25 of 26 antibodies able to neutralise the type 3 Sabin vaccine strain recognised a single antigenic site, distinct from that recognised by the remaining antibody.

The mutations were identified by limited sequencing of the genomic RNA of mutants and for the main site mutations were found to be clustered in the region of the genome encoding residues 89–100 of VP1 (MINOR et al. 1983, 1985; EVANS et al. 1983). The secondary site was also located in VP1 between residues 286 and 290 (MINOR et al. 1985). Two mutants in the major group had substitutions away from the main cluster in the primary sequence at residues 166 and 253 in VP1, respectively.

Similar studies were carried out on the type 1 Sabin and Mahoney strains, but the results obtained were strikingly different (EMINI et al. 1982; DIAMOND et al. 1985; MINOR et al. 1986a; PAGE et al. 1988). The number of mutational groups identified was greater, with no single group predominating as had been found for type 3, and mutations were located in a number of clusters none of which corresponded to the sites identified for type 3. Several of the sites were complex, having components from different proteins or different regions of the same protein. One site included sequences in VP2 from residues 164 and 170, sequences in VP1 from residues 221 to 226, and residue 270 of VP2. A second site included residues 58–60, 71 and 73 of VP3, and a third site residue 72 of VP2 and 76 of VP3. The striking difference in the antigenically significant mutational loci of type 1 and type 3 and the binding of the antibodies to synthetic peptides in ELISA led some to conclude that mutations rarely if ever occurred at the site to which the antibodies bound, and that effects were exerted on the conformation of distant epitopes (DIAMOND et al. 1985; BLONDEL et al. 1986).

When the mutational loci were identified on the surface of the virion it was clear that related mutations clustered into exposed distinct and prominent features and that it was unlikely that they could act by exerting a significant effect on distant conformations (HOGLE et al. 1985). All mutations in the principal site of type 3 clustered into a single feature at the pentameric apex of the virus, which included a loop made up of amino acids corresponding to residues 89–100, but also comprised three other loops, one of which included amino acid 166, and a second amino acid 253, which were implicated in the site by the mutational analysis. Moreover the mutations in type 1 in VP3 clustered in a region fairly close to the subsidiary site in VP1 of type 3 as one of the features around the pseudo-sixfold axis of symmetry. It was therefore concluded that the mutations were occurring at the site of attachment, and that at least one site was analogous to a site in type 3.

It is now apparent that the differences between the serotypes are due to major and unexpected differences in the relative immunogenicity of the different sites present on type 1 and type 3 for the mouse, with the result that the original monoclonal antibody panels did not encompass all sites.

BLONDEL et al. (1986) selected mutants from both the Sabin and Mahoney type 1 strains, using a panel including many of the same antibodies used by EMINI et al. (1983a) and identified mutations in VP3 at residues 60 and 71 in Mahoney and at residues 58 and 60 in the Sabin type 1 strain, and in VP1 at residue 100. The mutation at this second site, which corresponds to the major immunodominant locus of type 3, was selected by an antibody designated C3 which had been prepared from mice immunised with denatured virus (BLONDEL et al. 1983). Unlike most other monoclonal antibodies with neutralising activity for poliovirus, C3 reacts with isolated virion protein VP1, making it possible to identify the sequence recognised directly as described later.

While the region corresponding to the immunodominant site of type 3 was not strongly immunogenic in type 1 in any of the studies reviewed above, DERNICK and co-workers have isolated a large number of antibodies specific for type 1 poliovirus which select for mutations within this locus (UHLIG et al. 1983; WIEGERS and DERNICK 1987; WIEGERS et al. 1986, 1988). They have also demonstrated the involvement of the loop composed of residues 141–152 of VP1 in this site (WIEGERS et al. 1989). In some instances the antibodies were generated by immunisation with purified viral proteins and the splenocytes stimulated in vitro with infectious virus before fusion (WIEGERS et al. 1986) but this was not always the case (UHLIG et al. 1983). The success of this group in generating antibodies against this site of type 1 poliovirus by conventional means remains unexplained (WIEGERS et al. 1988). Antibodies against this site have also been generated by unusual immunisation routes, such as intrasplenic injection (MINOR et al. 1987).

Sites on type 3 virus other than the major immunodominant loop can be revealed by destroying its highly immunogenic character by treating the virus with trypsin, which introduces a single cleavage in the central portion of the immunodominant loop, largely destroying its antigenic and immunogenic character without affecting virus viability (FRICKS et al. 1985; ICENOGLE et al. 1986). Monoclonal antibodies prepared from mice immunised with trypsin cleaved type 3 poliovirus (MINOR et al. 1986a) selected mutations in VP2 at residues 164, 166, 167 and 172, in one group; in VP1 at residues 287 and 290 or in VP3 at residues 58, 59 in another and in VP3 at residues 77 and 79 in a third group. These sites correspond well with mutational loci identified in type 1.

While the differing findings with type 1 and type 3 poliovirus were reconciled to some extent by the studies described above and as summarised in Table 1, a number of features remain poorly explained. Firstly, it is unclear why the major loop at the pentameric apex of the virus is a strongly immunodominant site in type 3 but not in type 1. A possible explanation for this is suggested by the increased exposure of the loop in the structure of the Sabin type 3 strain (FILMAN et al. 1989) which is displaced by up to 8 Å compared with the type 1 loop and is therefore possibly more accessible to high affinity antibodies. Differences in dominance could be explained in terms of maturation of the immune response to favour such antibodies. This hypothesis remains speculative.

A second unexplained feature in the comparison of the structures of type 1 and type 3 is that while the site involving residues in VP2 from 164 to 170 is associated with a sequence in VP1 from 220 to 226 in type 1, the VP1 component

Table 1. Mutations in antigenic variants selected with monoclonal antibodies from poliovirus types 1, 2 and 3

Site	Mutated residue Type 1	Type 2	Type 3
1	VP1 97, 99, 100, 101 (1) VP1 100 (2) VP1 99 (3) VP1 144 (9)	VP1 94, 95, 97, 98, 99, 174, (4) VP1 93, 95, 96, 99 100, 101, 105 (5)	VP1 89, 91, 93, 95, 97, 98, 99, 100, 166, 253 (8)
2	VP1 221, 222, 223 (6) VP2 270 (6) VP1 221, 223, 224, 226 (7) VP2 164, 165, 165, 168, 169, 170 (7) VP2 270 (7) VP2 169, 170 (4) VP1 221, 223 (4)		VP2 164, 166, 167, 172 (4)
3	VP3 60, 71, 73 (6) VP3 58, 60, 71 (1) VP3 58, 59, 60 (7) VP3 58, 59, 60, 71 (4)		VP1 286, 287, 288, 290 (4, 8) VP3 58, 59 (4)
4	VP2 72 (6) VP2 72 (7) VP3 76 (7)		VP3 77, 79 (4)

Data from: (1) WIEGERS et al. (1988); (2) BLONDEL et al. 1986; (3) MINOR et al. (1987); (4) MINOR et al. (1986a); (5) LA MONICA et al. (1987); (6) DIAMOND et al. (1985); (7) PAGE et al. (1988); (8) MINOR et al. (1985); (9) WIEGERS et al. 1989

has not been demonstrated in type 3. Conversely, the sequence in VP1 from 286 to 290 has been shown to be antigenically significant in type 3, in association with VP3 58–60, while in type 1 only the VP3 component has been reported.

The antigenic structure of type 2 has not been studied as intensively as either type 1 or type 3. MINOR et al. (1986a) reported the isolation of mutants of the Sabin 2 strain resistant to type-specific antibodies. Of seven antibodies, five were directed against a site in VP1 from residues 94 to 99, corresponding to the immunodominant site of type 3. The two other antibodies failed to give mutants when used in selection, possibly because they recognise a site which cannot be changed without loss of viability. LA MONICA et al. (1987) used the same antibodies to select resistant mutants from the Lansing strain of type 2 poliovirus, which, unlike most strains of poliovirus, will kill mice by a poliomyelitis-like disease following intracerebral inoculation. The results obtained were generally consistent with those obtained with the Sabin type 2 strain, in that mutations were located in a sequence of VP1 from residues 93 to 105, and the two antibodies referred to above failed to give mutants.

However of 22 mutants selected, ten were less virulent than the parent when injected intracerebrally into mice. Four of the ten were also temperature sensitive in their growth in HeLa cells, and three of these had a deletion at residue 105. Reduction of neurovirulence in non-temperature sensitive viruses correlated with specific substitutions at residues 100 and 101. The interpretation of the results is complicated by the fact that of 14 mutants whose sequence was described, two had two mutations and one had three mutations within the sequence 91–105, raising the possibility of further mutations elsewhere in the genome. However, these findings indicate that mutations within an antigenic loop can affect the biological properties of the virus.

Residue 105 is believed to be located in the beta barrel structure of the virus, by analogy with the type 1 and type 3 structures, while the other substitutions which affect the antigenic structure are believed to be in the loop. The effect that deleting residue 105 has on the antigenic properties of the virus therefore implies that it is exerting an influence on a relatively distant conformation.

The sites which have been identified by isolation of antigenic variants in the type 1 strains Mahoney and Sabin, the type 2 strains Lansing and Sabin 2, and the type 3 strains Leon and Sabin 3 are shown in Table 1. Some monoclonal antibodies specific for site 1 have been shown to react with intact infectious virions and heated particles , and site 1 is thus expressed on particles having both N(D) and H(C) character (BLONDEL et al. 1983; FERGUSON et al. 1984). Sites 2 and 3 have been shown to be present on the intact infectious particles, and also on the 14S pentamers which assemble to form it (ROMBAUT et al. 1983; PAGE et al. 1988; EMINI et al. 1982) and site 4, which structurally is likely to form as a result of interactions between pentamers, has been found only on intact infectious virus or empty capsids expressing N(D) antigenic character and not on the 14S pentameric subunits (EMINI et al. 1982; DIAMOND et al. 1985; PAGE et al. 1988).

The conclusions drawn from mutational studies of the antigenic structure of poliovirus are consistent with the known atomic structure and with the properties of antigenic chimaeras produced by genetic engineering described in section 3.4 below. It remains the strategy of choice in many respects. However the results may be confused by effects of immunogenicity, which limit the available repertoire of monoclonal antibodies, the occurrence of mutations exerting effects on a distant conformation, or the inevitable selection of mutations in regions of an antigenic site which are best able to tolerate change without compromising virus viability. It is possible that such mutable regions are present in the virus specifically to evade the immune mechanisms of the host, as has been suggested for influenza (COLMAN et al. 1983).

3.3 Chemical, Immunochemical and Immunological Approaches to the Determination of the Antigenic Structure of Poliovirus

EMINI et al. (1982) studied two monoclonal antibodies, designated H3 and D3, which had neutralising activity for the type 1 poliovirus Mahoney. The bifunctional reagent toluene diisocyanate (TDI) was used to cross-link purified Fab fragments of the antibodies to the virus. The Fab fragments were linked to TDI at pH 7.2, then allowed to attach to the virus before the second functional group of the TDI was activated at pH 9.6, after which the virus was disrupted and the proteins separated by SDS-PAGE. The virion protein VP1 was extensively cross-linked, whereas, in contrast, exposure of EMC virus to the activated Fab fragment, or exposure of type 1 virus to TDI activated ovalbumin did not result in such extensive cross-linking of VP1. It was therefore concluded that both antibodies bound to VP1 of type 1.

It was later shown that while antibody H3 selected for mutations within VP1 at residues 221–223, antibody D3 selected for mutations in VP2 at residue 72. D3 was therefore directed against the only site on the virus in which VP1 has not been implicated as a contributing protein (see Table 1). Despite the care taken to minimise nonspecific binding, therefore, the results were almost certainly confused by linking of the antibodies to the most exposed virion protein in a way which was independent of their true site.

Alternative direct methods for the identification of antigenic determinants include the use of short synthetic peptides. If a short peptide takes up the conformation adopted by a similar sequence in the antigen it may be sufficiently antigenically active to bind monoclonal antibodies. EMINI et al. (1983c) identified a number of segments of the polio capsid proteins as possible antigenic sites based on assessment of hydrophilicity and sequence divergence. Monoclonal antibodies were reported to bind to peptides encompassing two of these sequences in ELISA. One, extending from residues 93 to 103 corresponded to site 1 of type 1, while the other, including residues 70–75 does not correspond to a known site and in fact is an internal sequence in the virion structure (HOGLE et al. 1985). The antibodies concerned were shown to select for mutants in entirely different sites (DIAMOND et al. 1985; BLONDEL et al. 1986) and reliance on the peptide binding data rather than the genetic approach lead to the misinterpretation of the results which was apparent when the structure was finally solved.

FERGUSON et al. (1986) examined the ability of monoclonal antibodies to bind to a peptide sequence derived from site 1 of type 3 poliovirus using an ELISA method. Antibodies specific for type 1 and type 2 poliovirus and influenza virus, which showed no reaction with type 3 virus, nonetheless bound more strongly to the peptide than antibodies specific for site 1 of type 3, indicating that in this case binding was nonspecific. Moreover monoclonal antibodies prepared against the peptide were shown to neutralise virus or to bind to peptide in ELISA, but not both, indicating that the major sequence conformation was different in the ELISA system to that in the virus itself.

An alternative method involving peptides uses them as immunogens. EMINI et al. (1983c) reported that peptide 93–103 described above could induce a neutralising response, where virus titre was reduced by $2 \log_{10}$ by incubation with an equal volume of undiluted serum. The reduction in titre when using sera raised against whole virus was greater than $8 \log_{10}$. CHOW et al. (1985) reported that a number of peptides would induce neutralising antibodies in rats detectable by sensitive assay methods. JAMESON et al. (1985a, b) reported the induction of neutralising antibodies by peptides, also using sensitive methods. In one case a standard neutralisation assay was performed (JAMESON et al. 1985a) and the dilution end point titre was 1 in 2; conventional hyperimmune rabbit anti-poliovirus sera would have titres in excess of 10 000 in such an assay. EMINI et al.

(1984) reported a sequence in VP2 which would induce a neutralising antibody response and corresponds to site 2. FERGUSON et al. (1985) hyperimmunised rabbits with a peptide corresponding to site 1 of type 3 poliovirus and generated an antiviral response in all rabbits, and a high neutralising titre of the order of 2500 in some. The sera induced failed to neutralise or bind to mutants with known substitutions within this site, and provided evidence that antibodies against this sequence were able to neutralise the virus. In general, however, the poliovirus-specific antibody titres generated by immunising animals with synthetic peptides have been either low or variable between individual animals.

The final method uses peptides to prime animals for an immune response to the virus (EMINI et al. 1983c, 1984). This supposes that the peptide bears a T cell epitope as well as an appropriate B cell epitope, and the experiments are technically difficult and laborious. It was reported (EMINI et al. 1985a) that a peptide derived from the sequence of poliovirus type 1 could prime for an immune response to hepatitis A, with which there was no sequence homology, implying that a peptide of inappropriate sequence can assume a conformation like that of a viral antigenic site. The sequence of the antigenic site is thus not necessarily the same as that of the peptide with priming activity, which makes the use of peptides to identify antigenic sites in this way uncertain.

Direct methods for identifying antigenic sites on poliovirus have therefore not been very successful, although two known antigenic sites have been identified or confirmed by their use. (EMINI et al. 1983b, 1984; FERGUSON et al. 1985). However, under circumstances where a monoclonal antibody recognises a linear epitope, as shown by reaction with isolated proteins in immunoprecipitation experiments or western blots, the site recognised can be identified convincingly using direct methods. For poliovirus such antibodies are exceptional, but two studies of this type have been reported.

The type 1 specific neutralising antibody C3 (BLONDEL et al. 1983, 1986; WYCHOWSKI et al. 1983; VAN DER WERF et al. 1983) was shown to precipitate isolated VP1. cDNA copies of fragments of the RNA coding for this region were inserted into an expression vector, and it was shown that only peptides expressing amino acids 90–104 of VP1 reacted with C3. Mutants selected with C3 had substitutions in VP1 at residue 100 (BLONDEL et al. 1986).

Monoclonal antibodies isolated by another group prepared by normal immunisation schedules (UHLIG et al. 1983) or in vivo priming followed by in vitro stimulation (WIEGERS et al. 1986) were shown to recognise linear epitopes within site 1 of poliovirus type 1 (WIEGERS and DERNICK 1987; WIEGERS et al. 1988). The methods included isolation of mutants, binding of antibodies to isolated proteins and competition with a peptide with the sequence of VP1 from residues 93 to 104. Of 56 monoclonal antibodies produced by in vitro stimulation, 49 reacted with VP1 and one with VP2 in Western blots, only five failing to react (WIEGERS and DERNICK 1987).

3.4 Antigenic Chimaeras of Polioviruses

The antigenic structure of poliovirus has been established largely by the isolation of nonneutralisable antigenic variants selected with monoclonal antibodies, whose properties have been shown to be consistent with the X-ray crystallographic structure.

A possible alternative way to identify components of antigenic sites is to use full-length infectious cDNA copies of poliovirus genomes and the available X-ray crystallographic structures to modify substantial sequences of the capsid proteins, and observe the effects on antigenicity, immunogenicity, viability and host range.

The first published description of an attempt to generate an antigenic chimaera of poliovirus having sequences of type 3 in a genome otherwise derived from type 1 was that of STANWAY et al. (1985) who replaced a sequence of VP1 of type 1 of 76 amino acids from residues 74 to 149 with the corresponding region of type 3, a process introducing a total of 14 amino acid differences. This work was carried out without the benefit of the detailed crystallographic structure. The resulting virus was reported to be nonviable, although recent studies suggest that this is not the case (K. BURKE, personal communication).

BURKE et al. (1988) and MURRAY et al. (1988a) reported more precise manipulations of the type 1 structure exploiting oligonucleotide directed mutagenesis of an infectious cDNA clone and a mutagenesis cartridge in an infectious cDNA clone respectively to replace an eight amino acid sequence of site 1 of type 1 with the corresponding sequence of type 3. The chimaeric virus was neutralised by polyclonal sera against both type 3 and type 1, and reacted with some, but not all type 3 monoclonal antibodies in neutralisation and antibody binding tests. Mice, rabbits and monkeys immunised with the chimaera and a monkey fed with the virus developed antibodies which would react with native, but not trypsin treated poliovirus type 3 in which the exchanged site was destroyed. The antibodies were thus not directed at a cross-reactive site present on both type 1 and type 3 (BURKE et al. 1988). The site was thus partially antigenically and immunogenically functional. The chimaeric virus did not grow as well as the parental Mahoney strain, giving smaller plaques and less than 10% of the virus yield at 37°C (MURRAY et al. 1988a).

COLBERE-GARAPIN et al. (1989) constructed chimaeras based on Sabin type 3, inserting tri- or hexapeptides and demonstrated the induction of antibodies with high specificity for the mutated viruses. Some of the mutants were of poor growth characteristics. These studies demonstrated that the loop of site 1 could be substantially extended by insertion without abolishing virus infectivity. The common observation that the chimaeras grew less well than the parental viruses suggested that the change of sequence has some functional effect, however.

Two further reports support this view (MURRAY et al. 1988b; MARTIN et al. 1988). In both, sequences from the mouse adapted type 2 virus Lansing were inserted into the Mahoney site 1 position using mutagenesis cartridges and resulting in changes in six amino acids. The viruses expressed joint type 1-type 2 antigenicity and immunogenicity, and also proved neurovirulent for mice in contrast to Mahoney itself. The ability of mutations in this sequence to attenuate Lansing for the mouse had been previously reported as described above (LA MONICA et al. 1987). While the construction of antigenic chimaeras is at an early stage, it will provide a means of studying the structure and function of the antigenic features of poliovirus in more detail than has been possible hitherto.

3.5 Natural Antigenic Variation in Polioviruses

Determination of the antigenic structure of poliovirus has relied heavily on the production of monoclonal antibodies in the mouse, which is not a natural host of the virus. Evidence for comparable sites in the natural human host is so far indirect. It has been known for some time that monoclonal and polyclonal antibodies are able to distinguish between strains of poliovirus (FERGUSON et al. 1982; VAN WEZEL and HAZENDONK 1979; CRAINIC et al. 1983; GUO et al. 1987), implying variation in antigenic sites between natural isolates from human hosts. Early studies with polyclonal sera and isolates from human recipients of the Sabin vaccine strains demonstrated that antigenic variation could occur during the period of excretion by vaccinees, especially in the case of type 1 strain (reviewed in WHO 1969). CRAINIC et al. (1983) reported that during excretion by a vaccinee, a Sabin-specific epitope could be lost and a Mahoney-specific epitope gained as defined by monoclonal antibody reactions, suggesting a response to antibody pressure in the gut. The epitopes were located in the site including residues 58–60 and 73 of VP3. Similar variation in this site for type 1 isolates was reported by MINOR et al. (1987).

JAMESON et al. (1985a) reported sequence variation in other regions of the Sabin type 1 strain during replication in the gut. Substitutions were identified in VP1 at residues 142 and 147, which is believed to form a component of the pentameric apical antigenic site (site 1 in Table 1). KEW and NOTTAY (1984) examined isolates which had been shown to be antigenically variant by their reactions with polyclonal sera by NAKANO et al. (1963). In two children there were substitutions in VP1 at residue 106, and in a third infant at residues 106 and 99. These results were tentatively interpreted to indicate that the virus was changing under antibody pressure, as the residues are located in a region corresponding to site 1. MINOR et al. (1986b) examined a series of sequential isolates of type 3 poliovirus from a primary vaccinee, and showed changes in antibody reaction after about 10 days with antibodies reacting with residues 160–170 of VP2 and a further change after 42 days with an antibody reacting with residues 77–79 of VP3. Changes in sequence consistent with the changes in reactivity were identified. The timing of the antigenic changes suggested that they arose in response to antibody pressure and the altered sites are not known to be involved in attenuation of the virus. Substitutions in these two antigenic sites have also been frequently found in isolates from other vaccinees (MINOR et al. 1989).

Collectively these findings imply that the human immune response to type 1 poliovirus recognises at least the VP3 strain specific site and possibly components of site 1 while the response to type 3 involves at least the sites located in VP2 between residues 160 and 170, and the site in VP3 at residue 77. ICENOGLE et al. (1986) also reported that the titre of serum from a human vaccinee was somewhat lower against trypsin-treated type 3 virus than the native form. The effect, which was far less marked than when mouse sera were studied, suggested that site 1 of type 3 might be recognised by the human immune response.

If this is the case it is surprising that antibodies against site 1 of type 3 are broadly reactive (FERGUSON et al. 1984; MINOR et al. 1987) and that the site is generally well conserved in sequence (MINOR et al. 1987) as an immunogenic site would be expected to vary under antibody pressure.

A possible explanation for this is provided by the fact that the site is sensitive to trypsin (ICENOGLE et al. 1986) and can therefore be shown to be cleaved and rendered antigenically inactive in its natural site of replication in the human gut (MINOR et al. 1987). A consequence of the presence of site 1 antibodies in the gut would be a selection pressure in favour of a trypsin cleavable site 1. Support for this hypothesis was provided by an outbreak of poliomyelitis in Finland, which was caused by an antigenically unusual type 3 strain which had lost the trypsin cleavage site in site 1, where it had an asparagine instead of the more usual arginine residue (HOVI et al. 1986; MAGRATH et al. 1986; HUOVILAINEN et al. 1987, 1988; MINOR et al. 1987). In several cases sequential isolates were obtained from infected individuals, and it was shown that over a period of about 2 weeks the asparagine residue mutated to a lysine residue by a single point mutation, thus introducing a trypsin cleavage site (MINOR et al. 1987; HUOVILAINEN et al. 1987).

The consensus sequence of wild type 1 isolates differs from that of wild type 3 isolates in that it has no trypsin cleavage site in antigenic site 1, and it is possible that, in humans, as in the mouse, it is not a potent immunogenic site. Attempts to raise human monoclonal antibodies have been reported, but the sites recognised remain to be determined. Some appear to be common to two or more serotypes (UYTDEHAAG et al. 1985; UHLIG and DERNICK 1988).

The trypsin cleavability of an antigenic site implies that type 3 poliovirus growing in the gut presents a significantly different antigenic structure to that growing elsewhere, or in a viraemia. This may serve to confine virus growth to the intestine where it is an essentially harmless infection. It also has possible significance for vaccination programmes which involve intact killed vaccine injected intramuscularly or live attenuated vaccine growing in the gut, as the two types of vaccine may stimulate different ranges of antibodies (ROIVANEN and HOVI 1987, 1988).

There is therefore indirect evidence that the human immune response recognises at least some of the antigenic structures defined by murine monoclonal antibodies. In view of the ease with which antigenic variants can be selected in vitro or in the human gut, the antigenic stability of polioviruses is surprising, and implies the presence of constraints on variation involving antigenic structure in some functional aspect, such as recognition of receptor sites. This is one possible implication of the change in host range of the Mahoney Lansing chimaeras.

4 Coxsackie B Viruses

Coxsackie B viruses are enteroviruses classified into six serotypes. They will infect and kill mice under suitable experimental conditions and have been implicated in a range of human diseases including diabetes and cardiac disorders. They are known to be capable of establishing a persistent infection in rodents and are believed to be able to do the same thing in human hosts, where it is possible that they produce chronic disease. The wide spectrum of clinical disease suggests variation between virus isolates. PRABHAKAR et al. (1982) reported that variants of coxcackie B4 could be selected at frequencies of up to one per 10^4 parental infectious virus units, and in natural isolates epitopes could be classified into highly, moderately or poorly conserved (PRABHAKAR et al. 1985). In these respects coxsackie B4 is virtually identical with the other picornaviruses. However, while all viruses were neutralised by hyperimmune sera, some epitopes recognised by monoclonal antibodies were lost on passage of the virus, and some gained without any known selection pressure, which has not been commonly reported for other picornaviruses and may imply a functional significance to the antigenic variation observed (but see BOLWELL et al. 1989a).

This view is supported by the properties of antigenic variants selected for resistance to monoclonal antibodies (PRABHAKAR et al. 1987). Mutants resistant to a single antibody were as virulent for suckling mice as the parent, killing all mice inoculated, while a double mutant killed only one-quarter and a triple mutant killed only 5%. In general resistance to the antibody was associated with failure of the antibody to react in radioimmune precipitation tests with virus, but for one antibody this was not the case. The sites recognised by the antibodies were not identified, and none of the antibodies bound to isolated proteins.

5 Hepatitis A

Hepatitis A is an atypical enterovirus in its sequence, extreme stability and probably its site of replication. There has been considerable interest in the virus because of its clinical importance but initially it proved difficult to cultivate so as to produce sufficient material for vaccine use. It was therefore felt to be a possible candidate for a subunit or peptide vaccine.

Direct methods similar to those used for poliovirus have been used to examine the antigenic structure of hepatitis A virus. They include immunisation of animals with isolated capsid proteins (HUGHES and STANTON 1985) or peptides (EMINI et al. 1985a, b; WHEELER et al. 1986), cross-linking of Fab fragments of monoclonal antibodies to virus with DTI (HUGHES et al. 1984) and the expression of genetically engineered virion proteins as beta galactosidase fusion products to probe serological responses (OSTERMAYR et al. 1987). These studies are subject to the reservations which have been summarised above for similar studies of poliovirus. In fact the extremely hardy nature of the virus, its tendency to aggregate and the difficulties inherent in its growth, purification and assay, make these studies even more difficult than those of poliovirus.

HUGHES et al. (1984) isolated a number of monoclonal antibodies specific for hepatitis A virus, and identified two groups of antibodies by competitive radioimmunoassay, where competition between members of different groups was minimal.

STAPLETON and LEMON (1987) used this method and the isolation of antigenic variants to demonstrate the presence of a dominant immunogenic site on the virus which may also be the major site recognised by human polyclonal sera. Sequencing studies indicated substitutions in residue 72 of VP3 and in VP1 at residue 102 (PING et al. 1988). The constancy and immunodominance of this antigenic site, such that mutants resistant to a single monoclonal antibody could also be resistant to human polyclonal sera, suggests that it may have some vital function. The antigenic variants proved unstable on passage in marmosets, reverting to the wild type, consistent with this view.

6 Rhinoviruses

6.1 Structure

The structure of rhinovirus 14 was solved to atomic resolution before that of the Mahoney strain of type 1 poliovirus (ROSSMANN et al. 1985) and played a significant part in clarifying the interpretation of the available data on the antigenic structure of picornaviruses. The structures of rhinovirus and poliovirus are very similar, as might be expected from the close relationship observed between their sequences, and differences are confined chiefly to the loops connecting the strands of the beta barrel of the three main structural proteins. The main loop from VP1 which constitutes the chief component of the immuno-dominant site of type 3 poliovirus is more exposed in rhinovirus 14 than it is in the Mahoney structure and the pentameric apex forms one of the most prominent features of the virus, as it does in polio. The internal protein VP4 is myristilated, but there is no lipid moiety inserted into the beta barrel of VP1 as is found in polio.

The canyon, which separates the pentameric apex from the face of the icosahedral particle, has been suggested to be the site through which the virus attaches to cells. Such a site is believed to be invariant since it has been shown that despite the fact that there are 100 serotypes of rhinovirus only two cell receptors are used. A mechanism for protecting the receptor site from immune pressure would be to conceal it in a cleft, so that antibodies might recognise residues around the rim, but would be too large to approach the virus attachment site itself.

6.2 Antigenic Structure of Rhinovirus

Rhinovirus 14 was studied by the isolation of antigenic mutants resistant to monoclonal antibodies (SHERRY and RUECKERT 1985; SHERRY et al. 1986). Four groups of mutants were identified by cross-neutralization studies suggesting the existence of at least four antigenic sites, termed neutralisation immunogens or Nims. The most commonly recognised site, NimIa, was the target of 15 of the 35 antibodies studied, and involved mutations in VP1 at residues 91 or 95. The next most common, NimIII was recognised by 11 antibodies, and involved residues 72, 75 and 78 in VP3, while NimII, recognised by five antibodies, involved residues 158, 159, 161 and 162 of VP2. The fourth site, NimIb, was recognised by four antibodies, and involved residues 83, 85, 138 and 139 of VP1. NimIa and NimIb were functionally distinct, despite their closeness in the sequence.

A number of anomalies were observed, including changes at residue 203 of VP3 or 287 of VP1 in NimIII. In the structure, these residues are close together. At least three mutants selected by one of the antibodies had more than one mutation, and the frequencies of isolation suggested that this antibody required a double substitution in a resistant mutant. Two of the mutants had substitutions in VP2 at both residues 161 and 162, and the third had a mutation in VP2 at residue 161, and a second in VP1 at residue 210.

The antigenic sites identified are structurally homologous to those of poliovirus, although there are some differences. Specifically the region corresponding to the single pentameric apical site of polio (site 1) forms two distinct sites, NimIa and NimIb, of which only the NimIa component has been demonstrated in poliovirus.

As for poliovirus the antigenic structure of rhinovirus 14 deduced by antigenic variants was consistent with the atomic structure. The data are summarised in Table 2.

Table 2. Mutations in antigenic variants selected with monoclonal antibodies from rhinovirus 14

Site	Mutated residues[a]
Nim 1a	VP1 91, 95
Nim 1b	VP1 83, 85, 138, 139
Nim II	VP2 158, 159, 161, 162; VP1 210
Nim III	VP3 72, 73, 75, 78, 203; VP1 287

[a] SHERRY et al. (1986)

6.3 Immunochemical and Immunological Approaches to the Determination of the Antigenic Structure of Rhinoviruses

Monoclonal antibodies which neutralise rhinoviruses in general do not recognise the isolated viral proteins, because of the strongly conformational and complex nature of the sites involved. However, SKERN et al. (1987) reported a monoclonal antibody raised by immunising mice with native rhinovirus 2, which was able to neutralise virus infectivity and to react in immunoblots with VP2. Regions of the genome encoding VP2 were inserted into an expression vector after introduction of limited deletions. It was found that the antibody would react with the expressed fragments provided they included the sequence of VP2 between residues 153 and 164 so identifying an antigenic site homologous to NimII of rhinovirus 14.

Studies with peptides have also been undertaken. FRANCIS et al. (1987a) examined the immunogenicity and antigenicity of a number of peptides whose sequence was derived from that of rhinovirus 2. One, corresponding to a region similar to that described by SKERN et al. (1987), induced a weak neutralising antibody response.

One of the difficulties in devising vaccines against rhinovirus is the large number of serotypes. The identification of a cross-protective epitope is therefore important in producing a vaccine against rhinovirus. McCRAY and WERNER (1987) immunised rabbits with a peptide representing sequences contributed by VP1 and VP3 to the canyon of rhinovirus 14; as described above this structure has been proposed to be the site by which the virion attaches to cells, and should therefore be invariant. Polyclonal antipeptide sera bound to HRV 14 in ELISA, and had weak neutralising activity against rhinovirus 14 and 27 other serotypes.

The levels of antibodies produced with peptides are comparable to those with poliovirus, and it is likely that the nature of most of the antigenic features of the virus will make it difficult to produce a potent vaccine based on a peptide.

7 Cardioviruses

7.1 Structure

The cardioviruses include Mengovirus and encephalomyocarditis virus (EMC) and for the purpose of this discussion Theiler's virus. Theiler's virus differs from the members of the group in that it lacks a polycytidilate tract in its genomic RNA and for a number of reasons has been classified as an enterovirus, but its sequence is 85% homologous to that of Mengovirus and EMC, and it shows very little homology with other enteroviruses.

Mengovirus was the third picornavirus whose structure was determined (LUO et al. 1987) and it is significantly different from both poliovirus and rhinovirus 14. The differences lie chiefly in the loops and extensions, the core structure of eight stranded beta barrels being similar to that of the other picornaviruses. The equivalent of the main immunogenic loop of VP1 of poliovirus and rhinovirus (site 1 and Nimla and 1b) which links the B and C strands of the VP1 beta barrel is greatly truncated in Mengovirus, and there is a large insertion forming two extensive loops linking the C strand and the D strand. These include residues 78–83 and 93–105. In VP2 a loop is deleted and a loop in VP3 is significantly displaced, but otherwise both proteins are arranged in a similar way to the corresponding proteins of polio and rhinovirus. The net effect of the differences seen is that the continuous depression or canyon found in the other two viruses around the pentameric apex is replaced by five distinct pores lined by sequences from VP1 rather than a mixture of sequences from VP1 and VP3. There is one main protrusion on the surface composed of sequences of VP1 from residues 58 to 64, 78 to 83 and 93 to 105 and of VP2 from residues 157 to 162. In addition at the C terminal 15 residues of VP1 are disordered but exposed.

7.2 Antigenic Structure of Cardioviruses and Theiler's Virus

LUND et al. (1977) reported that for Mengovirus isolated VP1 and VP2 but not VP3 were able to induce the production of neutralising antibodies. CLATCH et al. (1987) reported similar findings for Theiler's virus, where VP1 and VP2 could induce antiviral antibodies. Only VP1 was able to induce significant levels of neutralising antibody, however. Moreover VP3 was able to induce antibodies which recognised whole virus only in ELISA. The inference drawn was that adsorption of the virus to the ELISA plate might break up the structure such that VP3 was exposed. This distortion of viral structure on adsorption to ELISA plates without a capture antibody is a well-recognised phenomenon for foot and mouth disease virus (MCCULLOUGH et al. 1987) and probably happens for polio.

OHARA et al. (1988) described two monoclonal antibodies with neutralising activity for Theiler's virus which reacted in western blots with the virus protein VP1. Reaction was abolished if the virus was treated with trypsin, which introduces a cleavage close to the C terminus of VP1. Mutants selected with these monoclonal antibodies had mutations at residues 268 and 270 of VP1; VP1 is 274 amino acids in length in Theiler's virus, and if the structure is similar to that of Mengo virus this region would be in the disordered surface region of the protein. This may account for the reactivity of the antibodies in western blots.

Antigenic differences between closely related cardio-viruses may result in different pathogenicity, as appears to be the case for coxsackievirus. This has been suggested for Theiler's virus (NITAYAPHON et al. 1985) and for diabetogenic or nondiabetogenic isolates of EMC, which could be distinguished with monoclonal antibodies (YOON et al. 1988). However, the isolates of EMC differed by at least one spot in their T1 oligonucleotide maps, implying multiple differences. The monoclonal antibody panel available was able to differentiate between a number of isolates of EMC and Mengovirus, indicating that antigenic variation between strains occurs in this group of viruses as well as in others, although the cardioviruses are often regarded as monotypic.

8 Foot and Mouth Disease Virus

8.1 Structure

The atomic structure of foot and mouth disease virus was the last of the four genera to be solved (ACHARYA et al. 1989) and the particular strain examined was of serotype O_1 strain BFS1860. The serotypes have significantly different detailed structural features, including the presence of disulphide bridges between protomers in the O_1 but not the A_{10} serotypes, but the overall properties are likely to be comparable. The particle shell which is smooth in comparison with the other picornaviruses extends 100–150 Å from the particle centre, with some additional external disordered structure. There is no canyon or pit. While the core structures of the proteins are similar to the other picornaviruses, as might be expected, the loops between adjacent strands of the beta barel of VP1 are truncated, especially those contributing to the pentameric apex where poliovirus and rhinovirus have major antigenic sites. The sequence of VP1 from residues 133 to 158 forms a disordered loop giving the external disordered structure referred to above. Residues 210–213 are also disordered, and the 17 C terminal residues, 197–213, form an arm which runs in a clockwise direction along the virion surface from VP1, over VP3 to VP1 of the adjacent protomer where it is close to the disordered loop. The disordered regions include known antigenic sites, as will be discussed in detail below.

The loops in VP2 are also truncated. Residue 130 is linked by a disulphide bond to VP1 134 which is disordered; as a result residues 130 to 132 are faint in the structure. One-half of the residues of VP4 are disordered, but known to line the pore at the pentameric apex; all of the residues of VP3 were located. While the solution of the structure contributes greatly to understanding the antigenic structure of the virus, there is still a lack of consensus about some aspects.

8.2 Immunogenic Approaches to the Determination of the Antigenic Structure of Foot and Mouth Disease Virus

Foot and mouth disease virus treated with trypsin loses the ability to infect cells or to induce neutralising antibody, although the particles remain morphologically intact (CAVANAGH et al. 1977). The phenomenon varies with the strain of virus examined, so that for example O_6 is more affected by trypsin than C997 (ROWLANDS et al. 1971), but VP1 is selectively cleaved, and the degree of cleavage correlates with the biological effect. It is generally believed that the structural effect is restricted to the site of cleavage, which occurs in the disordered loop in the virion. STROHMAIER et al. (1982) exploited the immunogenicity of the VP1 of strain O_1K to locate sequences capable of inducing neutralising antibody, by fragmenting purified VP1 with cyanogen bromide, mouse submaxillary gland protease (MSGP) and trypsin. The fragments were identified by terminal sequencing and their immunogenicity studied to identify immunogenic sequences. Regions able to induce neutralising antibodies were found to lie between 146 and 154 and 201 and 213.

PFAFF et al. (1982) and BITTLE et al. (1982) used these findings, and theoretical consideration of hydrophilicity, sequence heterogeneity between strains and predicted structures of VP1 to identify sequences as probable antigenic sites which were then chemically synthesised as peptides. Both groups examined the O_1K strain, and reported that the sequence of VP1 from residues 144–160 (PFAFF et al. 1982) or from 141–160 (BITTLE et al. 1982) induced high levels of neutralising antibody. BITTLE also described a slightly less effective peptide representing residues 201–213, while PFAFF reported that a peptide representing residues 205–213 was ineffective. The two peptide sequences identified correspond to the two exposed regions of disordered sequence in the virion described above (ACHARYA et al. 1989), and it has been suggested that the disorder of the sequence contributes to the effective immunogenicity of the peptides compared to corresponding regions of other proteins and picornaviruses (WESTHOF et al. 1984; HARRISON 1989).

Peptides based on the sequences of the region corresponding to residues 151–160 of A_{10}, A_{12}, O_1, C3 and SAT2 strains of FMDV have been shown to induce neutralising antibodies, and the antibodies neutralise the homologous but not heterologous strains of serotype A which differ at 8 of the 20 residues (CLARKE et al. 1983).

The specificity of the response induced by peptides was further illustrated by ROWLANDS et al. (1983) who isolated four distinct strains designated A, B, C and USA from a single pool of A_{12}, and showed that they differed in the sequence from residues 141 to 168, as shown in Table 3. Sera raised against peptides based on the sequences from residues 141–160 of the four strains reflected the sequence relationships, such that A and C were closely related but distinct, as were B and USA. Minimal differences can therefore be detected by the peptide sera. This is comparable to later findings for poliovirus type 3 (FERGUSON et al. 1985). However, in the instance of the four strains of FMDV A_{12}, antisera raised with

Table 3. Sequences of residues 146–155 of VP1 strains of A_{12} virus from a single pool (ROWLANDS et al. 1983)

	146	147	148	149	150	151	152	153	154	155
A	Gly	Asp	Ser	Gly	Ser	Leu	Ala	Leu	Arg	Val
B			Leu					Pro		
C			Ser					Ser		
USA			Phe					Pro		

virions showed a specificity for the four strains similar to that found with antipeptide sera. While concentrated sera were able to neutralise all four strains, these findings strongly indicate that the major part of the antibody response to the A12 strain is directed against a single antigenic loop. An issue which has not yet been resolved, but which will be discussed below, is whether the immune response involves antibodies against a number of entirely distinct sites, or whether it is directed exclusively to an antigenic site composed of this loop either alone or in conjunction with other parts of the virus.

The induction of neutralising antibodies which protect guinea pigs (BITTLE et al 1982) or cattle and swine (KLEID et al. 1981) raises the possibility of subunit or synthetic vaccines, thus eliminating the risk of infecting animals with inadequately inactivated virus vaccines and greatly simplifying production. Consequently attention has been paid to identifying protective epitopes (MELOEN and BARTELING 1986a) and peptides protective in the target species which do not require a carrier protein (DiMARCHI et al. 1986; DOEL et al. 1988). A significant effort has been directed towards improving the response generated, including studies of immunological priming (FRANCIS et al. 1985, 1988) and the effect of extending the peptide with known T cell epitopes (FRANCIS et al. 1987b) and alternative modes of presentations such as adding the sequences to other immunogens. This has included their insertion into one of the antigenic sites of the influenza haemagglutinin, which was then expressed in vaccinia (NEWTON et al. 1987), their addition to the N terminus of the hepatitis B core protein then expressed in vaccinia (NEWTON et al. 1987; CLARKE et al. 1987), or their expression as single, double or quadruple copies at the N terminus of beta galactosidase (BROEKHUIJSEN et al. 1987). The beta galactosidase fusion protein was shown to protect the natural host, and the hepatitis B core construct in particular produced a dramatic protective immune response, approaching that found with conventional vaccines.

8.3 Immunochemical Approaches to the Determination of the Antigenic Structure of Foot and Mouth Disease Virus

As for the other picornaviruses monoclonal antibodies can be prepared against FMDV which recognise intact virus, subviral particles or isolated virion proteins. Some will recognise only intact virions implying strong conformational elements in some of the antigenic sites recognised (OULDRIDGE et al. 1984; STAVE et al. 1986, 1988) but some will react with all three types of preparation. DUCHESNE et al. (1984) reported an antibody which would react in Western blots and also neutralised, and was therefore similar in character to the poliovirus specific monoclonal antibody studied by WYCHOWSKI et al. (1983) and the rhinovirus 2 specific antibody studied by SKERN et al. (1987).

Their antibody would react with VP1, but not if it had previously been treated with either trypsin or MSGP, which cleave VP1 at known sites (STROHMAIER et al. 1982). It has concluded that the epitope recognised included residues 146 and 147 of VP1. STAVE et al. (1986) reported an antibody of similar type, reacting with isolated VP1 of the O_1 strain in a solid phase radioimmunoassay. By examining the reactions of expressed truncated fragments of VP1, it was shown that the epitope recognised lay between residues 137 and 172. The epitopes identified by these two studies were thus consistent with those identified by the use of immunogenic peptides.

McCULLOUGH et al. (1987) reported studies of the antigenic structure of the O_1 Suisse strain of FMDV by competition ELISA and identified six sites.

Binding of monoclonal antibodies to synthetic peptides has been used extensively to identify the sites which they recognise. GEYSEN et al. (1984) reported a method for synthesising small quantities of peptide immobilised on polythene rods, rather than on column resins. This greatly simplifies the mechanics of synthesis and the assessment of reactions in ELISA, and made possible the synthesis of every overlapping hexapeptide in the sequence of VP1 of foot and mouth (i.e. residues 1–6, 2–7, 3–8 etc.) which were then tested for reaction with appropriate sera. The assumptions on which this approach is based include the following (a) the antigenic sites involved are linear epitopes of continuous sequence of six amino acid residues or less; (b) the correct conformation will be adopted by a sufficient amount of the peptide to make it recognisable; (c) binding of antibody to peptide is absolutely specific. Experience with poliovirus makes each of these assumptions highly questionable, but the two main antigenic loops of FMDV from residues 146–152 and 206–212 in VP1 were identified as the principal targets of the antibodies in a polyclonal serum.

ROBERTSON et al. (1984) mapped monoclonal antibodies specific for the A12 strain using peptides as antigens. MELOEN and BARTELING (1968b) reported studies of overlapping hexapeptides of VP1 from the O_1, A_{10}, and C1 strains with sera raised with virus, subviral particles or VP1, and reported that peptides from the same regions were active for each virus. MELOEN et al. (1987) examined two neutralising antibodies specific for the A_{10} strain for their reaction with overlapping hexapeptides of VP1. An octapeptide deduced from these studies was as effective as virus in binding the monoclonal antibody. PFAFF et al. (1988) examined the O_1K strain using four monoclonal antibodies, two of which reacted only with whole virus, while two others reacted with virus and isolated VP1 and could be shown to recognise the 141–160 loop of VP1 by binding to peptides in ELISA or by competition for binding to the virus by peptide in solution.

Finally PARRY and co-workers (1985, 1989) have described the use of competition between peptides in solution and immobilised virus in ELISA to identify the epitopes recognised by monoclonal antibodies. It is possible to obtain an indication of the avidity of the antibody for the peptide by measuring ED50 levels, and thus to distinguish specific and nonspecific binding. PARRY et al. (1989) described studies with four antibodies and a series of peptides and concluded that all four recognised two components of VP1; one recognised residues 146–150 and 200–213, two recognised 143–146 and 200–213 and the fourth recognised residues 161–170 and 200–213. In each case sera raised against the individual peptides would compete with the monoclonal antibody for binding to virus, but only to 50%. Full inhibition required a mixture of sera raised against each site. This result is somewhat unexpected in that an antibody reacting with one-half of the epitope recognised would be predicted to prevent the antibody binding completely or reduce its net avidity without affecting the maximum binding. An explanation based on affinities was suggested. The earlier paper (PARRY et al. 1985) was the first to suggest that the sequences from residues 141–160 and 200–213 could form part of the same site and thus be close together in the structure, as proved to be the case.

The results obtained with peptides suggest the existence of only two linked sites. They are not in complete agreement with those obtained by the use of genetic methods, however, which suggests that other loci may affect antigenic properties.

8.4 Genetic Approaches to the Determination of the Antigenic Structure of Foot and Mouth Disease Virus

Several studies of the O serotype have been carried out in which antigenic variants have been isolated and characterised. In most studies mutations have been identified in the known immunogenic loop in VP1, consistent with the reactions of the antibodies with peptides representing this region and its known immunodominance. However, all studies have also reported mutations outside this site and the antigenic reactions of the mutants currently imply the existence of other independent sites. In one study antibodies reacting with O_1 Kaufbeuren were examined by the isolation of antigenic variants and reaction in ELISA with the parental virus (XIE et al. 1987; MCCAHON et al. 1989; J.D.A. KITSON, D. MCCAHON and G.J. BELSHAM, in preparation). The data strongly suggest that the panel of antibodies recognised a total of four independent antigenic sites (see Sect. 3.2).

One group included antibodies which bound to viral subunits, VP1 and peptides representing the sequence of VP1 from residues 141–160. In agreement with this the mutants generated by these antibodies could be shown to have substitutions at residues 144, 154 and 208 of VP1, which individually conferred partial resistance, or at residue 148, which conferred complete resistance. Mutations at residues 152 and 171 have also been found in association with a substitution at 144 (KITSON et al., in preparation).

The other groups of antibodies recognised conformational epitopes only present on intact virus. Mutations in VP2 were identified in one group at residues 70, 71, 72, 73, 75, 77 and 131. In one instance a double mutant was selected having a mutation in VP2 at residue 72 and in the VP1 site above at residue 144 (KITSON et al., in preparation). Its antigenic properties reflected that it was a double mutant. The third group selected for mutations in VP1 at residues 43, 44 and 45 and the fourth group selected for mutants in VP3 at residue 58.

STAVE et al. (1988) identified three groups of monoclonal antibodies with neutralising activity for the O_1B strain. Group I reacted only with intact virions, group II with virions and 12S subunits and group III with intact virions, 12S subunits and VP1. Group III included an antibody shown by reaction with expressed fragments of VP1 to recognise a sequence of VP1 between residues 135 and 172 (STAVE et al. 1986, see above). Mutants selected with this antibody, and another of the same group had amino acid substitutions at residue 148 or at 138 and 144 in agreement with the results obtained using the direct method. Sequence changes for mutants of the other groups were not identified. The antigenic reactivity of mutants selected with the different antibody groups suggested that while group I was unique, groups II and III overlapped.

BARNETT et al. (1989) selected mutants from the O₁ BFS strain. Three groups were identified. Nine of 11 antibodies were in one group and the epitopes recognised were present on virions, trypsin-treated virions and subviral particles. Mutants selected with two of these antibodies showed changes in the isoelectric point of capsid protein VP2, and seven out of nine of the mutants sequenced had a substitution at residue 134 of VP2. Findings with the other mutants of this group were not reported.

One of the remaining two antibodies was in a unique group and four of 11 mutants examined had substitutions at amino acid 43 of VP1, while a fifth had a substitution at amino acid 48 of VP1. The remaining antibody failed to give mutants.

PFAFF and co-workers (1988) described antibodies which recognised conformational epitopes on intact virus. Mutants resistant to one of them were found to have multiple substitutions in the capsid region, including changes at residues 200 and 207 (corresponding to VP2 residues 130 and 137) and at residues 435, 436 and 460, (corresponding to VP3 residues 146, 147 and 171). The large number of differences reported makes the results difficult to interpret. A summary of the findings with type O is given in Table 4.

Serotype A differs from O in a number of respects including the absence of a disulphide bridge between VP2 and VP1 of the adjacent protomer (ACHARYA

Table 4. Mutations in antigenic variants selected with monoclonal antibodies from foot and mouth disease virus type O

Group	Mutated residue	Reference
1	VP1 144, 148, 152, 154, 208, 171	XIE et al. 1987; McCAHON et al. 1989
2	VP2 70, 71, 72, 73, 75, 77, 131	
3	VP1 43, 44, 45	
4	VP3 58	
1	VP1 138, 144, 148	STAVE et al. 1988
2	not known	
3	not known	
1	VP2 134, not known	BARNETT et al. 1989
2	VP1 43, 48	
3	not known	
1[a]	VP2 130, 137,	PFAFF et al. 1988
	VP3 146, 147, 171	

[a] All substitutions found in one mutant

et al. 1989) and the A_{10} subtype was studied by THOMAS et al. (1989). The extent to which the different mutational loci involved separate independently mutable sites was not fully established as information on cross-neutralisation was not presented on each mutant isolated, nor was the complete sequence of the capsid region determined, although multiple mutants were described. The data were consistent with either a single site or several entirely independent sites. The resolution of this question is important in view of the possible different interpretations of the genetic studies discussed below. However, the data indicate that the region in VP1 from residues 142 to 153 is a major locus, consistent with studies of immunogenic peptides and that a second major locus lies in VP3 in a position analogous to group 4 of the O_1 subtype studied by XIE et al. (1987) and MacCAHON et al. (1989). The locus in VP3 is complex, including residues 58–70, 136, 195 and residue 80 of VP2, where substitutions appeared in association with residues 70 and 196 of VP3. The other two regions identified involved residue 169 of VP1 and residue 204 of VP1.

BOLWELL et al. (1989b) examined ten antibodies specific for a strain of the A serotype. The mutants isolated fell into three groups, which related well to the sites identified by reaction with peptides bound to ELISA plates. Group 1 gave mutations at residues 149, 150, 153 and 154 of VP1, group 2, which overlapped with group 1, gave mutants with substitutions at residues 138, 140, 142 and 144, and group 3 gave mutants where substitutions were not identified. The results are summarised in Table 5. These findings are compatible with those obtained by THOMAS et al. (1988) and with the O_1 strains by others.

There is no consensus on the interpretation of the genetic studies of the antigenic structure of foot and mouth disease virus at present. At least three views can be taken. Firstly, by analogy with the other picornaviruses and other viruses studied by these methods (COLMAN et al. 1983) it is possible that all mutations occur at the site to which the antibody binds. For foot and mouth disease virus this is accepted as the case where antibodies have been shown to bind to peptides with sequences derived from that of VP1 between residues 140 and 160 and to be

Table 5. Mutations in antigenic variants selected with monoclonal antibodies from foot and mouth disease type A

Group	Mutated residue	Reference
1	VP1 142, 144, 146, 147, 149 150, 151, 152, 153, 157*	THOMAS et al. (1988)
2	VP1 204	
3	VP3 58, 59, 61, 70, 136*, 139, 195 VP2 80, 133*, 196*	
4	VP1 169	
1	VP1 149, 150, 153, 154	BOLWELL et al. 1989b
2	VP1 138, 140, 142, 144	
3	not known	
1	VP2 82*, 88*, 207*	BOLWELL et al. 1989a

*Substitution found only in multiple mutant

affected by mutations in the virus in the same region (XIE et al. 1987; McCAHON et al. 1989; STAVE et al. 1988; THOMAS et al. 1988; BOLWELL et al. 1989b).

However, mutational loci have been identified in other proteins which affect a large proportion of antibodies (see Tables 4 and 5), and in view of the strong immunodominance of the trypsin-sensitive site on the virus and the observed reactions of antipeptide sera, it has been proposed that mutations outside the immunodominant loops may be exerting their effect by modifying a distant conformation (BARNETT et al. 1989).

A third possibility is that at least some of the antibodies bind to the immunodominant loop and to the residues in the mutated region but that the mutated region is best able to accomodate the required changes. The antibodies selecting for mutations in residues 130–135 of VP2 may be an example of this, although no group has yet reported a monoclonal antibody which is equally affected by mutations in the immunodominant loops and in VP2. However, PFAFF, et al. (1988) described an antibody which clearly bound to the VP1 sequence but gave rise to mutants which had multiple substitutions including two in VP2 at residues 130 and 137 but no substitutions in VP1. Secondly, BARNETT et al. (1989) reported that the majority of their antibodies fell into a single group, and that mutants selected by two of these antibodies had substitutions in VP2 at residue 134. The immunodominant locus would be expected to be in VP1, and PARRY et al. (1989) describe peptide-binding data consistent with the view that this group of antibodies in fact bind to this locus. THOMAS et al. (1988) reported mutations in the immunodominant loop in a multiple mutant which also had a substitution in VP2 at residue 132. Finally, X-ray crystallographic studies indicate that residue 130 of VP2 is close to residue 134 of VP1, to which it is linked by a disulphide bridge. Residue 134 of VP1 is at the start of the disordered sequence forming the immunodominant site. Similar considerations may apply to the other loci.

Which, if any, of these interpretations is correct is not yet established.

8.5 Functional Significance of Antigenic Sites of FMDV

There is evidence implicating the antigenic sites of FMDV in viral attachment. BOLWELL et al. (1989a) described the selection of different antigenic variants of an A22 strain of FMDV by passage on BHK cells in suspension or in monolayer. Monoclonal antibodies raised against virus grown in monolayer reacted with a number of different A isolates, while those raised against virus grown in suspension were specific for that virus. The two types of virus differed in their adsorption to cells in monolayer, and in the sequence of VP2 at residues 82, 88 and 207, but not in VP1.

The immunodominant loop of FMDV contains the sequence arginine-glycine-aspartate (RGD), found in the cell adhesion protein fibronectin. In FMDV it is highly conserved between a great range of isolates, although it lies in a hypervariable region (Fox et al. 1989). There is no canyon or pit in the FMDV

structure to function as a possible cell attachment site (ACHARYA et al. 1989) and
it was shown that synthetic peptide containing the RGD sequence, including one
based on the sequence of Sindbis virus, and antisera directed against them would
prevent attachment of FMDV to cells, while corresponding peptides without it
would not. It was therefore suggested that this sequence is responsible for
attachment of the virus to cells, and is protected from immune surveillance by its
relatively small size and location in a hypervariable region.

9 Conclusions

The antigenic structure of the picornaviruses has been studied in some detail
using a variety of methods, including synthetic peptides and the isolation of
antigenic mutants, and the interpretation of the results has been greatly aided by
the availability of the three-dimensional structure of at least one member of each
of the main groups. The effectiveness of the approaches used has been largely
determined by the structural nature of the sites concerned. Thus, disordered sites,
such as the main FMDV loop and possibly the C terminus of VP1 of Theiler's
virus have been readily imitated by synthetic peptides. Complex ordered
structures, such as site 3 of polio (Table 1) have not been imitated by such
peptides and their identification depends on the isolation of antigenic mutants. It
is possible that a loop of fixed conformation can be imitated by peptides, provided
it is free standing, so that its conformation is determined solely by its primary
amino acid sequence rather than its context in the antigen. No loop in
picornaviruses is entirely free standing in this sense. While a consensus on the
structure of the more intensively studied viruses has developed in general, some
issues clearly remain to be resolved. In particular the possible biological functions
of regions of the virus which are antigenically significant are beginning to be
explored.

References

Acharya R, Fry E, Stuart D, Fox G, Rowlands DJ, Brown F (1989) The three dimensional structure of
    foot and mouth disease virus at 2.9 A resolution. Nature 338: 709–716
Barnett PV, Ouldridge EJ, Rowlands DJ, Brown F, Parry NR (1989) Neutralizing epitopes of type O
    foot and mouth disease virus I. Identification and characterization of three functionally
    independent conformational sites. J Gen Virol 70: 1483–1492
Beatrice ST, Katze MG, Zajac BA, Crowe RL (1980) Induction of neutralizing antibodies by the
    coxsackievirus B3 viron polypeptide, VP2. Virology 104: 426–438
Bittle JL, Houghten RA, Alexander H, Shinnick TM, Sutcliff JG, Lemen RA, Rowlands DJ, Brown F
    (1982) Protection against foot and mouth disease by immunisation with a chemically synthesised
    peptide from the viral nucleotide sequence. Nature 298: 30–33

Blondel B, Crainic R, Horodniceanu F (1982) Le polypeptide structural VP1 du poliovirus induit des anticorps neutralizants. CR Acd Sci (III) 294: 91–94

Blondel B, Akachem O, Crainic R, Couillin P, Horodniceanu F (1983) Detection by monoclonal antibodies of an antigenic determinant critical for poliovirus neutralization present on VP and heat inactivated virions. Virology 126: 707–710

Blondel B, Crainic R, Fichet O, Dufraisse G, Cardnea A, Diamond D, Girard M, Horaud F (1986) Mutations conferring resistance to neutralization with monoclonal antibodies in type 1 poliovirus can be located outside or inside the antibody binding site. J Virol 57: 81–90

Bolwell C, Brown AL, Barnett PV, Campbell RO, Clarke BE, Parry NR, Ouldridge EJ, Brown F, Rowlands DJ (1989a) Host cell selection of antigenic variants of foot and mouth disease virus. J Gen Virol 70: 45–57

Bolwell C, Clarke BE, Parry NR, Ouldridge EJ, Brown F, Rowlands DJ (1989b) Epitope mapping of foot and mouth disease virus with neutralizing monoclonal antibodies. J Gen Virol 70: 59–68

Breindl M (1971) VP4, the D reactive part of poliovirus. Virology 46: 962–964

Broekhuijsen MP, Van Rijn JMM, Blom AJM, Pouwels PH, Enger-Valk BE, Brown F, Francies MJ (1987) Fusion proteins with multiple copies of the major antigenic determinant of foot and mouth disease virus protect both the natural host and laboratory animals. J Gen Virol 68: 3137–3143

Burke K, Dunn G, Ferguson M, Minor PD, Almond JW (1988) Antigenic chimeras of poliovirus: potential novel vaccines against picornaviral infections. Nature 332: 81–82

Cavanagh D, Sangar DV, Rowlands DJ, Brown F (1977) Immunogenic and cell attachment sites of FMDV: further evidence for their location in a single capsid polypeptide. J Gen Virol 35: 149–158

Chow M, Baltimore D (1982) Isolated capsid protein VP1 induces a neutralizing response in rats. Proc Natl Acad Sci USA 79: 7518–7512

Chow M, Yabrov R, Bittle J, Hogle J, Baltimore D (1985) Synthetic peptides from four separate regions of the poliovirus type 1 capsid protein VP1 induce neutralizing antibodies. Proc Natl Acad Sci USA 82: 910–914

Chow M, Newman JFE, Filman D, Hogle JM, Rowlands DJ, Brown F (1987) Myristylation of picornavirus capsid protein VP4 and its structural significance. Nature 327: 482–486

Clarke BE, Casrroll AR, Rowlands DJ, Nicholson BH, Houghten RA, Lemen RA, Brown F (1983) Synthetic peptides mimic subtype specificity of foot and mouth disease virus. FEBS Lett 157: 261–264

Clarke BE, Newton SE, Carroll AR, Francis MJ, Appleyard G, Syned AD, Highfield PE, Rowlands DJ, Brown F (1987) Improved immunogenicity of a peptide epitope after fusion to hepatitis B core protein. Nature 330: 381–384

Clatch RJ, Pevear DC, Rozton E, Roos RP, Miller SD, Lipton HL (1987) Characterisation and of humoral immune responses to Theiler's murine specificity encephalomyelitis virus capsid proteins. J Gen Virol 68: 3191–3196

Colbere-Garapin F, Christodoulous C, Crainic R, Garapin A–C, Candrea A (1989) Addition of a foreign to the major capsid protein of poliovirus oligopeptide. Proc Nat Acad Sci USA 85: 8668–8672

Colman PM, Vanghese JN, Laver WC (1983) Structure of the catalytic and antigenic sites in influenza virus neuraminidase. Nature 303: 41–44

Crainic R, Couillin P, Blondel B, Cabau N, Boue A, Horodniceanu F (1983) Natural variation of poliovirus neutralization epitopes. Infect Immun 41: 1217–1225

Dernick R, Heukeshjoven J, Hilbrig M (1983) Induction of neutralizing antibodies by all three structural poliovirus polypeptides. Virology 130: 243–246

Diamond DC, Jameson BA, Bonin J, Kohara M, Abe S, Itoh H, Komatsu T, Arita M, Kuge S, Nomoto A, Osterhaus ADME, Crainic R, Wimmer E (1985) Antigenic variation and resistance to neutralization in poliovirus type 1. Science 229: 1090–1093

Di Marchi R, Brooke G, Gale C, Cracknell V, Doel T, Mowat N (1986) Protection of cattle against foot and mouth disease by a synthetic peptide. Science 232: 639–641

Doel T, Gale C, Brooke G, Di Marchi R (1988) Immunisation against foot and mouth disease with synthetic peptide representing the C terminal region of VP1. J Gen Virol 69: 2403–2406

Duchesne M, Cartwright T, Crespo A, Boncher F, Fallound A (1984) Localization of a neutralization epitope of foot and mouth disease virus using neutralizing monoclonal antibodies. J Gen Virol 65: 1559–1566

Emini EA, Jameson BA, Lewis AJ, Larsen GR, Wimmer E (1982) Poliovirus neutralization epitopes: analysis and localisation with neutralizing monoclonal antibodies. J Virol 43: 997–1005

Emini EA, Dorner AJ, Dorner LF, Jameson BA, Wimmer E (1983a) Identification of a poliovirus neutralization epitope through use of neutralizing antiserum raised against a purified viral structural protein. Virology 124: 144–151

Emini EA, Kao SY, Lewis AJ, Crainic R, Wimmer E (1983b) Functional basis of poliovirus neutralization determined with monospecific neutralizing antibodies. J Virol 46: 466–474

Emini EA, Jameson BA, Wimmer E (1983c) Priming for and induction of antipoliovirus neutralizing antibodies by synthetic peptides. Nature 304: 699–703

Emini EA, Jameson BA, Wimmer E (1984) Identification of a new neutralizing antigenic site on poliovirus coat protein VP2. J Virol 52: 719–721

Emini EA, Berger J, Hughes JV, Miba SW, Linemeyer DL (1985a) Priming of an anti hepatitis A virus antibody response by poliovirus-specific synthetic peptides: localization of potential antigenic sites of hepatitis A virus neutralization. In: Lerner RA, Chennock RH, Brown F (eds) Vaccines 85. Molecular and chemical basis of resistance to parasitic bacterial and viral diseases. Cold Spring Harbor Laboratory, Cold Spring Harbor 217–220

Emini EA, Hughes JV, Perlow DS, Boger J (1985b) Induction of hepatitis A virus neutralizing antibody by a virus specific synthetic peptide. J Virol 55: 836–839

Evans DM, Minor PD, Schild GC, Almond JW (1983) Critical role of an eight amino acid sequence of VP1 in neutralization of poliovirus type 3. Nature 304: 439–462

Ferguson M, Qi Yi-Hua, Minor PD, Magrath DI, Spitz M, Schild GC (1982) Monoclonal antibodies specific for the Sabin vaccine strain of poliovirus 3. Lancet ii: 122–124

Spitz M, Schild GC (1982) Monoclonal antibodies specific for the Sabin vaccine strain of poliovirus 3. Lancet ii: 22–124

Ferguson M, Minor PD, Magrath DI, Qi Yi-Hua, Spitz M, Schild GC (1984) Neutralization epitopes on poliovirus type 3 particles: analysis using monoclonal antibodies. J Gen Virol 65: 197–201

Ferguson M, Evans DMA, Magrath DI, Minor PD, Almond JW, Schild GC (1985) Induction of broadly reactive, type specific neutralizing antibody to poliovirus type 3 by synthetic peptides. Virology 143: 505–515

Ferguson M, Reed SE, Minor PD (1986) Reactivity of anti-peptide and anti-poliovirus type 3 monoclonal antibodies with synthetic peptides. J Gen Virol 67: 2527–2531

Filman DJ, Syed R, Chow M, Macadam AJ, Minor PD, Hogle JM (1989) Structural factors that control conformational transitions and serotype specificity in type 3 poliovirus. EMBO J 8: 1567–1579

Fox G, Parry NR, Barnett PV, McGinn B, Rowlands DJ, Brown F (1989) The cell attachment site on foot and mouth disease virus includes the amino acid sequence RGD (arginine-glycine-aspartic acid). J Gen Virol 70: 625–637

Francis MJ, Fry CM, Rowlands DJ, Brown F, Bittle JL, Houghten RA, Lerner RA (1985) Immunological priming with synthetic peptides of foot and mouth disease virus. J Gen Virol 66: 2347–2354

Francis MJ, Hastings GZ, Sangar DV, Clark RP, Syred A, Clarke BE, Rowlands DJ, Brown F (1987a) A synthetic peptide which elicits neutralizing antibody against rhinovirus type 2. J Gen Virol 68: 2687–2691

Francis MJ, Hastings GZ, Syred AD, McGinn B, Brown F, Rowlands DJ (1987b) Nonresponsiveness to a foot and mouth disease virus peptide overcome by addition of foreign helper T-cell determinants. Nature 300: 168–170

Francis MJ, Fry CM, Rowlands DJ, Brown F (1988) qualitative and quantitative differences in the immune response to foot and mouth disease virus antigens and synthetic peptides. J Gen Virol 69: 2483–2491

Fricks CE, Icenogle JP, Hogle JM (1985) Trypsin sensitivity of the Sabin strain of type 1 poliovirus: cleavage sites in virions and related particles. J Virol 54: 856–859

Fromhagen LL (1965) The separation and physicochemical properties of the C and D antigens of coxsackievirus. J Immunol 95: 818–822

Geysen HM, Meloen RH, Barteling SJ (1984) Use of peptide synthesis to probe viral antigens for epitopes to a resolution of a single amino acid. Proc Natl Acad Sci USA 81: 3998–4002

Guo R, Tang E-H, Wang H, Yang X-F, Liu M-Y, Qin H-X, Li QH, Zhuang J-Y, Liu K-M (1987) Preliminary studies on antigenic variation of poliovirus using neutralizing monoclonal antibodies. J Gen Virol 68: 989–994

Harrison SC (1989) Picornaviruses: finding the receptors. Nature 338: 205–206

Harrison SC, Olson AJ, Schutt CE, Winkler FK, Bricogne G (1978) Tomato bushy stunt virus at 2.9 Å resolution. Nature 276: 368–373

Hogle JM, Chow M, Filman DJ (1985) Three dimensional structure of poliovirus at 2.9 Å resolution. Science 229: 1358–1365

Hopp TP, Woods KR (1981) Prediction of protein antigenic determinants from amino acid sequence. Proc Natl Acad Sci USA 78: 3824–3828

Hovi T, Cantell H, Hovilainen A, Kinnunen E, Kuronen T, Laninleimu K, Pogry T, Roivairen M, Saloma N, Stenvik M, Silander A, Theden C-J, Salminen S, Weckstrom P (1986) Outbreak of paralytic poliomyelitis in Finland: widespread circulation of antigenically altered poliovirus type 3 in a vaccinated population. Lancet ii: 1427–1432

Hughes JV, Stanton LW (1985) Isolation and immunization with hepatitis A viral structural proteins: induction of antiprotein, antiviral and neutralizing responses. J Virol 55: 395–401

Hughes JV, Stanton LW, Thomassini JE, Long WJ, Scolnick EM (1984) Neutralizing monoclonal antibodies to hepatitis A virus: parallel localization of neutralizing antigenic site. J Virol 52: 465–473

Hummeler K, Tumilowicz JS (1960) Studies on the complement fixing antigens of poliomyelitis II. Preparation of type specific anti-N and anti-H indicator sera. J Immunol 84: 630–634

Huovilainen A, Hovi T, Kunnunen L Takkinen K, Ferguson M, Minor PD (1987) Evolution of poliovirus during an outbreak: sequential type 3 poliovirus isolates from several persons show shifts of neutralization determinants. J Gen Virol 68: 1373–1378

Houvilainen A, Kunnunen L, Ferguson M, Hovi T, (1988) Antigenic variation among 173 strains of type 3 poliovirus isolated in Finland during the 1984 to 1985 outbreak. J Gen Virol 69: 1941–1948

Icenogle JP, Minor PD, Ferguson M, Hogle JM (1986) Modulation of the humoral response to a 12 amino acid site on the poliovirus virion. J Virol 60: 297–301

Jameson BA, Bonin J, Wimmer E, Kew OM (1985a) Natural variants of the Sabin type vaccine strain of poliovirus and correlation with a poliovirus neutralization site. Virology 143: 337–341

Jameson BS, Bonin J, Murray MG, Wimmer E, Kew OM (1985b) Peptide induced neutralizing antibodies to poliovirus. In: Lerner AR, Channock RM, Brown F (eds) Vaccines 85. Molecular and chemical basis of resistance to parasitic, bacterial and viral diseases. Spring Harbor Laboratory pp 191–198

Joklik WL, Darnell JE (1961) The adsorption and early fate of purified poliovirus in HeLa cells. Virology 13: 439–447

Kew OM, Nottay B (1984) Evolution of the oral poliovaccine strains in human occurs by both mutation and intramolecular recombination. In: Chanock RM, Lerner RA (eds) Modern approaches to vaccines 1: molecular and chemical basis of virus virulence and immunogenicity. Cold Spring Harbor Laboratory, Cold Spring Harbor, pp 357–362

Kleid DG, Yansura D, Small B, Dowbenko D, Moore DM, Grubman MJ, McKercher PD, Morgan DO, Robertsosn BH, Bachrach HL (1981) Cloned viral protein vaccine for foot and mouth disease: responses in cattle and swine. Science 214: 1125–1129

La Monica N, Kupsky WJ, Racaniello VR (1987) Reduced neurovirulence of poliovirus type 2 Lansing antigenic variants selected with monoclonal antibodies. Virology 161: 429–437

Le Bouvier GL (1955) The modification of poliovirus antigens by heat and ultraviolet light. Lancet ii: 1013–1016

Lonberg Holm K, Butterworth BE (1976) Investigation of the structure of polio and human rhinovirions through the use of selective chemical reactivity. Virology 71: 207–216

Lonberg Holm K, Yin FH (1983) Antigenic determinants of infective and inactivated hyman rhinovirus type 2. J Virol 12: 114–123

Lonberg Holm K, Gosser LB, Kauer JC (1975) Early alteration of poliovirus in infected cells and its specific inhibition. J Gen Virol 27: 329–342

Lund GA, Ziola BK, Salmi A, Scraba VG (1977) Structure of a Mengo virion. V. Distribution of the capsid polypeptides with respect to the surface of the virus particle. Virology 78: 35–44

Luo M, Vriend G, Kamer G, Minor I, Arnold E, Rossmann MG, Boege U, Scraba DG, Duke GM, Palmenbert AC (1987) The atomic structure of Mengovirus at 3.0 Å resolution. Science 235: 182–191

Magrath DI, Evans DMA, Ferguson M, Schild GC, Minor PD, Horaud F, Crainic R, Stenvik M, Hovi T, (1986) Antigenic and molecular properties of type 3 poliovirus responsible for an outbreak of poliomyelitis in a vaccinated population. J Gen virol 67: 899–905

Martin A, Wychowski C, Conderc T, Crainic K, Hogle J, Giraud M (1988) Engineering a poliovirus type 2 antigenic site on a type 1 capsid results in a chimaeric virus which is neurovirulent for mice. EMBO J 7: 2839: 2847

Mayer MM, Rapp HJ, Roizman B, Klein SW, Cowan KM, Lukens D, Schwerck CE, Schaffer FL, Charney J (1957) The purification of poliomyelitis virus as studied by complement fixation. J Immunol 78: 435–455

McCahon D, Crowther JR, Belsham GJ, Kitson JDA, Duchesne M, Have P, Meloen RH,

Morgan DO, de Simone F (1989) Evidence for at least four antigenic sites on type O foot and mouth disease virus involved in neutralisation: identification by single and multiple site monoclonal antibody resistant mutants. J Gen Virol 70: 639–645

McCray J, Werner G (1987) Different rhinovirus serotypes neutralized by antipeptide antibodies. Nature 329: 736–738

McCullough KC, Crowther J-R, Carpenter WC, Brachi E, Capucci L, de Simone F, Xie Q, McCahon D (1987) Epitopes on foot and mouth disease virus particles. Virology 157: 516–525

Meloen RH, Barteling SJ (1986a) An epitope located at the C terminus of isolated VP1 of foot and mouth disease virus type O induces neutralising activity but poor protection. J Gen Virol 67: 289–294

Meloen RH, Barteling SJ (1986b) Epitope mapping of the outer structural protein VP1 of three different serotype of foot and mouth disease virus. Virology 149: 55–63

Meloen RH, Rowlands DJ, Brown F (1979) Comparison of the antibodies elicited by hte individual structural polypeptides of foot and mouth disease and polioviruses. J Gen Virol 45: 761: 763

Meloen RH, Puyk WC, Meijer DJA, Lankhof H, Posthumus WPA, Schaaper WMM (1987) Antigenicity and immunogenicity of synthetic peptides of foot and mouth disease virus. J Gen Virol 68: 305–314

Minor PD (1987) Structure of picornavirus coat proteins and their antigenicity In: Rowlands DJ, Mayo M, Mahy BWJ (eds) The molecular biology of the positive strand RNA viruses. Academic, New York, pp 259–280 (FEMS Symposium, Vol 32)

Minor PD, Schild GC, Bootman J, Evans DMA, Ferguson M, Reeve P, Spitz M, Stanway F, Cann AJ, Hauptmann R, Clarke L-D, Mountford, RC, Almond JW (1983) Location and primary structure of a major antigenic site for poliovirus neutralization. Nature 301: 674–679

Minor PD, Evans DMA, Ferguson M, Schild GC, Westrop G, Almond JW (1985) Principal and subsidiary antigenic sites of VP1 involved in the neutralization of poliovirus type 3. J Gen Virol 65: 1159–1165

Minor PD, Ferguson M, Evans DMA, Almond JW, Icenogle JP (1986a) Antigenic structure of polioviruses of serotypes 1, 2 and 3. J Gen Virol 67: 1283–1291

Icenogle JP (1986a) Antigenic structure of polioviruses of serotypes 1, 2 and 3. J Gen Virol 67: 1283–1291

Minor PD, John A, Ferguson M, Icenogle, JP (1986b) Antigenic and molecular evolution of the vaccine strain of type 3 poliovirus during the period of excretion by a primary vaccinee. J Gen Virol 67: 693–706

Minor PD, Ferguson M, Phillips A, Magrath DI, Huovilainen A, Hovi T (1987) Conservation in vivo of protease cleavage sites in antigenic sites of poliovirus. J Gen Virol 68: 1857–1865

Minor PD, Dunn G, Evans DMA, Magrath DI, John A, Howlett J, Phillips A, Westrop G, Wareham K, Almond JW, Hogle JM (1989) The temperature sensitivity of the Sabin type 3 vaccine strain of polioviruses: molecular and structural effects of mutation in the capsid protein VP3. J Gen Virol 70: 1117–1123

Murray MG, Kuhn RJ, Arita M, Kawamura N, Nomoto A, Wimmer E (1988a) Poliovirus type 1/type 3 antigenic hybrid virus constructed in vitro elicits type 1 and type 3 neutralizing antibodies in rabbits and monkeys. Proc Natl Acad Sci USA 85: 3203–3207

Murray MG, Bradley J, Yang X-F, Wimmer E, Moss EG, Racaniello VR (1988b) Poliovirus host range is determined by a short amino acid sequence in neutralization antigenic site 1. Science 241: 213–215

Nakano JH, Gelfand HM, Cole JT (1963) The use of a modified Wecker technique for the serodifferentiation of type 1 polioviruses related and unrelated to Sabin's vaccine strains II. Antigenic segregation of isolates from specimens collected in field studies. Am J Hyg 78: 215–230

Newton SE, Clarke BE, Appleyard G, Francis MJ, Carroll AR, Rowlands DJ, Skehel J, Brown F (1987) New approaches to FMDV antigen presentation using vaccinia virus. In: Brown F, Chanock RM, Lerner RA (eds) Vaccines 87. Cold Spring Harbor Laboratory, Cold Spring Harbor, pp 12–21

Nitayaphon S, Toth MM, Ross RP (1985) Localization of a neutralization site of Theiler's murine encephalomyelitis viruses. J Virol 56: 887–895

Ohara Y, Senkowski A, Fu J, Klaman L, Goodall J, Toth M, Ross RP (1988) Trypsin-sensitive neutralization site on VP1 of Theiler's murine encephalomyelitis viruses. J Virol 62: 3527–3529

Ostermayr R, Von der Helm K, Gauss-Muller V, Winnacker EL, Deinhardt F (1987) Expression of heptatitis A virus cDNA in Eschericia coli: antigenic VP1 recombinant protein. J Virol 61: 3645–3647

Ouldridge EJ, Barnett PV, Parry N-K, Syred A, Head M, Kneyemamu MM (1984) Demonstration of neutralizing and nonneutralizing epitopes on the trypsin-sensitive site of foot and mouth disease virus. J Gen Virol 65: 203–207

Page GS, Moser AG, Hogle JM, Filman DJ, Rueckert RR, Chow M (1988) Three dimensional structure of poliovirus serotype 1 neutralizing determinants. J Virol 62: 1781–1794

Parry NR, Ouldridge EJ, Barnett PV, Rowlands DJ, Brown F, Bittle JL, Houghten RA, Lerner RA (1985) Identification of neutralizing epitopes of foot and mouth disease virus. IN: Lerner RA, Channock RM, Brown F (eds) Cold Spring Harbor Laboratory, Cold Spring Harbor, pp 211–216

Parry NR, Barnett PV, Ouldridge EJ, Rowlands DJ, Brown F (1989) Neutralizing epitopes of type O foot and mouth disease virus II: mapping three conformational sites with synthetic peptide reagents. J Gen Virol 70: 1493–1503

Pfaff E, Mussgay M, Bohm HO, Schulz GE, Schaller H (1982) Antibodies against a preselected peptide recognise and neutralise foot and mouth disease virus. EMBO J 1: 869–874

Pfaff E, Thiel H-J, Beck E, Strohmaier K, Schaller H (1988) Analysis of neutralizing epitopes on foot and mouth disease virus. J Virus 62: 2033–2040

Ring L-H, Jansen RW, Stapleton JT, Cohen JI, Lemon SM (1988) Identification of an immunodominant antigenic site involving the capsid protein VP3 of hepatitis A virus. Proc. Natl Acad Sci USA 85: 8281–8285

Prabhakar BS, Harpel MV, McClintock PR, Notkins AL (1982) High frequency of antigenic variants among naturally occurring human coxsackie B4 virus isolates identified by monoclonal antibodies. Nature 300: 374–376

Prabhakar BS, Menegus MA, Notkins AL (1985) Detection of conserved and nonconserved epitopes on coxsackievirus B4: frequency of antigenic change. Virology 146: 302–306

Prabhakar BS, Srinivasappa J, Ray V (1987) Selection of coxsackie B4 variants with monoclonal antibodies results in attenuation. J Gen Virol 68: 865–869

Robertson BH, Morgan DO, Moore DM (1984) Location of a neutralizing epitopes defined by monoclonal antibodies generated against the outer capsid polypeptide BP1 of foot and mouth disease A12. Virus Res 1: 489–500

Roivanen M, Hovi T (1987) Intestinal trypsin can significantly modify antigenic properties of polioviruses: implications for the use of inactivated poliovirus vaccine. J Virol 61: 3749–3754

Roivanen M, Hovi T (1988) Cleavage of VP1 and modification of antigenic 1 site of type 2 polioviruses by intestinal trypsin. J Virol 62: 3536–3539

Rombaut B, Vrijsen R, Boeye A (1983) Epitope evolution in poliovirus maturation. Arch Virol 76: 289–298

Rosmann MG, Arnold A, Erickson JW, Frankenberger EA, Griffith JP, Hecht H-J, Johnson JE, Kamer G, Luo M, Mosser AG, Rueckert RR, Sherry B, Vriend G (1985) Structure of a human common cold virus and functional relationship to other picornaviruses. Nature 317: 145–153

Rowlands DJ, Sangar DV, Brown F (1971) Relationship of the antigenic structure of foot and mouth disease virus to the process of infection. J Gen Virol 13: 85–93

Rowlands DJ, Sangar DV, Brown F (1975) A comparative chemical and serological study of the full and empty particles of foot and mouth disease virus. J Gen virol 26: 227–238

Rowlands DJ, Clarke BE, Carroll AR, Brown F, Nicholson BH, Bittle JL, Hoghten RA, Lerner RA (1983) Chemical basis of antigenic variation in foot and mouth disease virus. Nature 306: 694–697

Rueckert RR, Wimmer E (1985) Systematic nomenclature of picornavirus proteins. J Virol 50: 957–959

Rweyemamu MM, Hingley PJ (1984) Foot and mouth disease virus strain differentiation: analysis of the serological data. J Biol Stand 12: 323–337

Sherry B, Rueckert R (1985) Evidence for at least two dominant neutralization antigens on human rhinovirus 14. J Virol 53: 137–143

Sherry B, Mosser AG, Colonno RJ, Rueckert RR (1986) Use of monoclonal antibodies to identify four neutralization immunogens on a common cold picornavirus human rhinovirus 14. J Virol 57: 246–257

Skern T, Neubauer C, Frasel L, Grundler P, Summergruber W, Zorn M, Kuechler E, Blass D (1987) A neutralizing epitope on human rhinovirus type 2 includes amino acid residues between 153 and 164 of virus capsid protein VP2. J Gen Virol 68: 315–323

Stanway G, Hughes PJ, Westrop GD, Evans DMA, Dunn G, Minor PD, Schild GC, Almond JW (1985) Construction of poliovirus intertypic recombination by use of cDNA. J Virol 57: 1187–1190

Stapleton J, Lemon S (1987) Neutralization escape mutants define a dominant immunogenic neutralization site on hepatitis A virus. J Virol 61: 491–498

Stave JW, Card JL, Morgan DO (1986) Analysis of feet and mouth disease virus type 01 Brugge neutralization epitopes using monoclonal antibodies. J Gen Virol 67: 2083–2092

Stave JW, Card JL, Morgan DO, Vakharia VN (1988) Neutralization sites of type 01 foot and mouth disease virus defined by monoclonal antibodies and neutralization escape virus variants. Virology 162: 21–29

Strohmaier K, Franze R, Adam KH (1982) Location and characterisation of the antigenic portion of the FMDV immunising protein. J Gen Virol 59: 295–306

Thomas AAM, Woortmeijer RJ, Puijk W, Barteling SJ (1988) Antigenic sites on foot and mouth disease virus type A10. J Virol 62: 2782–2789

Uhlig H, Dernick R (1988) Intertypic cross-neutralization of polioviruses by human monoclonal antibodies. Virology 163: 214–217

Uhlig H, Rutter G, Dernick R (1983) Evidence for several unrelated neutralisation epitopes of poliovirus type 1, strain Mahoney, provided by neutralization tests and quantitative enzyme linked immunosorbent assay (ELISA). J Gen Virol 64: 2809–2812

Uytdehaag FGCM, Loggen HG, Logtenberg T, Lichtveld RA, van Steenis B, van Asten JAAM, Osterhause ADME (1985) Human monoclonal antibodies to cross reactive epitopes of poliovirus. J Immunol 135: 3094–3101

van der Marel P, Hazenbonk TG, Henneke MAC, van Wezel AL (1983) Induction of neutralizing antibodies by poliovirus capsid polypeptides VP1, VP2 and VP3. Vaccine 1: 17–22

van der Werf S, Wychowski C, Bruneau P, Blondel B, Crainic R, Horodinceanu F, Girard M (1983) Localization of a poliovirus type 1 neutralization epitope in viral capsid polypeptide VP1. Proc Natl Acad Sci USA 80: 5080–5084

van Wezel AL, Hazendonk AG (1979) Intratypic differentiation of poliomyelitis virus strains by strain-specific antisera. Intervirology 11: 2–8

Westhof E, Altschuh D, Moras D, Bloomer AC, Mondragon A, Klug A, van Regenmortel MHV (1984) Correlation between segmental mobility and the location of antigenic determinants in proteins. Nature 311: 123–126

Wheeler CM, Robertson BH, van Nest G, Dina D, bradley DW, Fields HA (1986) Structure of the hepatitis A virion: peptide mapping of the capsid region. J Virol 58: 307–313

WHO (1969) Evidence on the safety and efficacy of live poliomyelitis vaccines currently in use, with special reference to type 3 poliovirus. Bull H O 40: 925–945 (WHO Memorandum)

Wiegers KJ, Dernick R (1987) Binding site of neutralizing monoclonal antibodies obtained after in vivo priming with purified VP1 of poliovirus type 1 is located between amino acid residues 93 and 104 of VP1. Virology 157: 248–251

Wiegers K, Uhlig H, Dernick R (1986) In vivo stimulation of presensitized mouse spleen cells with poliovirus type 1, Mahoney, and enhancement of poliovirus specific hybridomas. J Gen Virol 67: 2053–2057

Wiegers K, Uhlig H, Dernick R (1988) Evidence for a complex structure of neutralization antigenic site of poliovirus type Mahoney. J Virol 62: 1845–1848

Wiegers K, Uhlig H, Dernick R (1989) N-Ag1B of poliovirus type 1: a discontinuous epitope formed by two loops of VP1 comprising residues 96–104 and 141–152. Virology 70: 583–586

Wychowski C, van der Werf S, Siffert O, Crainic R, Bruneau P, Girard M (1983) A poliovirus type 1 neutralization epitope is located within amino acid residues 93–104 on viral capsid polypeptide. EMBO 2: 2019–2024

Xie Q-C, McCahon D, Crowther JR, Belsham GJ, McCullough KC (1987) Neutralization of foot and mouth disease virus can be mediated through any of at least three separate antigenic sites. J Gen Virol 68: 1637–1647

Yoon J-W, Kaw Bae Y-S, Pak CY, Amano K, Eun H-M, Kim MK (1988) Identification of antigenic differences between the diabetogenic and non-diabetogenic variants of encephalomyocarditis virus using monoclonal antibodies. J Gen Virol 69: 1085–1090

Poliovirus Genetics

P. Sarnow, S. J. Jacobson, and L. Najita

1 Introduction

The goal of genetic studies of animal viruses has been the elucidation of the structure and function of the viral genome. The ultimate aim has been the assignment of functions to individual viral polypeptides in infected host cells and the understanding of genetic processes such as transmission, recombination, and complementation. Necessary tools to accomplish these goals are viral mutants, defective in particular processes, whose lesions can be assigned to particular regions of the viral genome.

Poliovirus genetics was born with the introduction of a plaque assay for poliovirus, in which one could assay individual progeny from mixed infections (Dulbecco 1954). This allowed the isolation of clonal strains of polioviruses and

Department of Biochemistry, Biophysics and Genetics, and Department of Microbiology and Immunology, University of Colorado Health Sciences Center, Denver, Colorado 80262, USA
* Part of this study was supported by Public Service grants AI-25105 and AG-07347 from the National Institutes of Health

provided an assay for viral mutants whose phenotypes differed from wild-type virus.

Subsequently, many poliovirus mutants were isolated (COOPER 1969, 1977) but have been only of limited use in the understanding of the function of the viral genome. This was mainly due to a lack of mutants which demonstrably bore single genetic lesions.

Poliovirus genetics enjoyed a renaissance with two discoveries. First, the entire nucleotide sequence of the viral genome was determined (KITAMURA et al. 1981; RACANIELLO and BALTIMORE 1981a) from which the structure and function of the viral genome could be predicted. Second, a cDNA copy of the viral RNA was found to yield infectious virus upon introduction into mammalian tissue culture cells (RACANIELLO and BALTIMORE 1981b), creating the possibility of obtaining viral mutants by genetically manipulating the cDNA at predetermined positions. The genetic lesions of such defined mutants can be mapped unambiguously and the infectious cDNA molecules can be used as the source of genetically homogeneous viral stocks.

In this chapter we would like to (a) review the new methods available for the genetic analysis of poliovirus; (b) summarize the information obtained on the structure and function of the viral genome with the aid of well-defined viral mutants; and (c) discuss genetic processes of poliovirus in infected host cells.

2 Genetic Approaches for Studying the Poliovirus Genome

2.1 Classical Genetic Approaches

Classical poliovirus genetics was directed towards the production of a genetic map of the viral genome. To accomplish this, a variety of viral mutants were obtained and used as genetic markers in recombination and complementation tests. Genetic markers most commonly used were temperature sensitive (ts) or drug-resistant viruses (COOPER 1969, 1977). Ts mutants were obtained by growing wild-type virus in the presence of the chemical mutagens 5-fluorouracil or nitrosoguanidine. Alternatively, infectious virion RNA was treated with nitrous acid or hydroxylamine. Both experimental approaches yielded ts mutants which produced 100- to 1000-fold less virus in a single infectious cycle at the nonpermissive temperature than wild-type virus did under similar conditions. The physiological properties of the mutants were examined by measuring the production of single- and double-stranded viral RNA molecules and the formation of infectious virions during a single infectious cycle.

Using 19 ts mutants in recombinant crosses, COOPER (1977) was able to assign the loci for structural and nonstructural proteins to opposite ends of the viral genome. He concluded that there are three primary gene functions encoded by the viral RNA: structural proteins, replicase I activity (an RNA polymerase activity which copies the viral genome to yield a negative strand RNA), and a replicase II activity (an RNA polymerase activity which copies the negative strands to yield positive strands).

The 5' to 3' orientation of genetic markers on a single-stranded genome can not be deduced simply by using recombination tests. Therefore, a biochemical approach, pactamycin mapping, was utilized to reveal the 5' to 3' orientation of the polioviral genome. Pactamycin inhibits the initiation of translation, but allows ribosomes to elongate preinitiated polypeptide chains. If pactamycin and a radioactive amino acid are added to poliovirus-infected cells, only those proteins whose translation was initiated prior to the addition of the drug become radioactively labeled. Therefore, proteins translated from the 3' end contain more label than proteins synthesized from the 5' end of the mRNA. In this way it was determined that structural viral proteins are translated from the 5' end and nonstructural proteins from the 3' end of RNA, and that all viral proteins arise from a single translational initiation event.

Classical poliovirus genetics has revealed that genetic recombination can occur on the RNA level, and can be used to produce a rough genetic map. In addition, it was suggested that genetic complementation outside the capsid region was limited (COOPER 1969, 1977). It now seems probable that the failure to detect genetic complementation between any ts mutants obtained by chemical mutagenesis was most likely due to multiple genetic lesions in the mutant RNA genomes. However, as will be discussed later, puzzling failures to complement among defined viral mutants are still observed (Sect. 5.2).

2.2 New Genetic Approaches

2.2.1 Isolation of an Infectious cDNA Copy of the Viral Genome

The entire nucleotide sequence of the poliovirus Mahoney type 1 RNA genome was reported in 1981 (KITAMURA et al. 1981; RACANIELLO and BALTIMORE 1981a). KITAMURA and colleagues sequenced the viral RNA using a modification of Sanger's chain termination method while RACANIELLO and BALTIMORE sequenced a cloned cDNA copy of the viral genome. The nucleotide sequence confirmed the existence of one large open reading frame encoding the viral polyprotein. Subsequently, 12 viral polypeptides, mapped by amino acid sequence analysis, were found to be proteolytic cleavage products of the polyprotein (KITAMURA et al. 1981).

A complete cDNA copy of the poliovirus genome was cloned into the vector pBR322, resulting in the 11.8 kilobase plasmid pVR106 (RACANIELLO and BALTIMORE 1981b). Transfection of pVR106 into human HeLa cells or monkey kidney CV1 cells resulted in the production of infectious virus which could be neutralized with rabbit antisera to poliovirus. Why did pVR106 give rise to infectious virus? It is thought that the plasmid migrated into the nucleus of the cell where cellular RNA polymerase II could use plasmid sequences which resemble

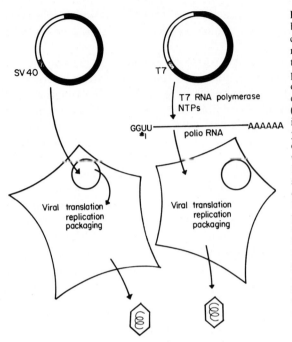

Fig. 1. Production of poliovirus by transfection of infectious cDNA and RNA molecules into mammalian cells. Plasmids containing the full-length cDNA of poliovirus (*solid bar*) downstream of a eukaryotic promoter (for example the SV40 early promoter; (*stippled bar*) can be transfected into appropriate primate cells to yield virus. It is thought that the cDNA migrates into the nucleus where it is transcribed by cellular RNA polymerase II. The transcript is then transported into the cytoplasm where it is translated by host ribosomes. Alternatively, plasmids containing the viral cDNA downstream of a bacteriophage promoter (the T7 ϕ10 promoter element; *stippled bar*) can be transcribed in vitro by T7 RNA polymerase to yield highly infectious RNA molecules which can be transfected directly into the cytoplasm of the host cell

eukaryotic promoters to transcribe the cDNA. Alternatively, the plasmid may have integrated into cellular chromosomes from where transcription could have been initiated by a neighboring cellular promoter element. In any case, the cloning of eukaryotic promoter elements in front of poliovirus cDNA sequences increased virus production 50- to 100-fold (SEMLER et al. 1984). However, the yield of virus obtained from transfected cDNAs is still very poor, even when eukaryotic shuttle vectors containing Simian virus 40 enhancer and promoter sequences are used. Using such vectors, 2000 plaque forming units (pfu) per μg transfected DNA could be obtained (SEMLER et al. 1984) compared with 10^6 pfu per μg transfected virion RNA (SPECTOR and BALTIMORE 1974). The low specific infectivity of transfected cDNAs is probably due to the prerequisite for the cDNA to migrate into the nucleus in order to be expressed (see Fig. 1); the primary polymerase II transcript most likely contains extraneous nucleotide sequences at its 5' and 3' ends. Also, the viral RNA contains fortuitous cryptic splicing signals which may cause splicing of many of the RNA molecules in the nucleus and therefore render the RNA noninfectious. In addition, any viral RNA remaining intact has to be transported from the nucleus into the cytoplasm where translation can initiate. All these situations are unusual for the viral RNA, because normally the viral life cycle takes place solely within the cytoplasm.

It has been known since 1958 that virion RNA is infectious (ALEXANDER et al. 1958). With the cloning of the promoter for T7 bacteriophage RNA polymerases (MCALLISTER et al. 1981) and of the gene for bacteriophage T7 RNA polymerase by DAVANLOO and co-workers (1984), it became possible to transcribe almost any cDNA in vitro. In order to synthesize polioviral RNA in vitro, VAN DER WERF and co-workers (1986) constructed a plasmid containing a full-length poliovirus cDNA downstream of the T7 ϕ promoter element (see Fig. 1). Upon addition of T7 RNA polymerase and nucleotide triphosphates (NTPs), microgram quantities of viral RNA could be synthesized in vitro. These RNA molecules contained two extraneous guanosine residues at their 5' end and heterogeneous 3' ends and were 5% as infectious as virion RNA when transfected into HeLa cells. Removal of the extraneous, nonvirally encoded 3' end sequences and addition of a polyadenosine tail resulted in viral RNA molecules which were as infectious as virion RNA (SARNOW 1989). Therefore, infectious poliovirus RNA can be synthesized in vitro and can be used (a) as a template for in vitro translation and replication studies and (b) as a way of introducing lethal mutants, made by site-directed mutagenesis (see below), into mammalian cells and hopefully to study their phenotypes in a single infectious cycle. Previously, such an approach had been difficult due to the low infectivities of cDNA clones.

2.2.2 Mutagenesis Methods

Prior to the availability of an infectious poliovirus cDNA, mutants were obtained primarily through chemical mutagenesis or by isolation of spontaneously arising mutants with a selectable phenotype (COOPER 1969, 1977). Recombinational mapping could then be performed to map the site of the mutation at a crude level (COOPER 1977). The major problem with this method of genetic analysis is that multiple mutations in the RNA are probably present in any mutant genome. Therefore, it is very difficult to correlate a particular genetic alteration with an observed phenotype. Considering the estimated error rate of the viral RNA polymerase to be approximately 10^{-3} to 10^{-4} errors per nucleotide incorporated (STEINHAUER and HOLLAND 1987), such mutants are relatively unstable; this genetic instability is manifested as a high reversion frequency. For these reasons, it is desirable to obtain well-defined viral mutants with only a single genetic alteration and a stable phenotype. Technological refinements in the manipulation of DNA have yielded a variety of methods by which mutations can be constructed. Two general mutagenesis schemes have been used to study the structure and function of the poliovirus RNA genome.

Site-directed mutagenesis is employed to introduce a predetermined lesion into the viral cDNA with subsequent analysis of the mutant phenotype. The selection of target sites for mutagenesis is usually aided by the identification of regions with sequence conservation among picornaviruses or regions of sequence homology to proteins with known function. Specific questions can then be asked about functional domains of the viral RNA or particular viral proteins. An advantage of this method is that one possesses the infectious plasmid harboring the mutation as a pure clone which helps in the further analysis of the introduced mutation (see below). A major drawback of this method is that most mutations in the cDNA do not give rise to mutants with a viable or conditional lethal phenotype. Although these cDNAs might harbor interesting mutations, the inability to recover virus renders them useless.

Random mutagenesis is designed to generate a large number of mutations covering the entire infectious cDNA. Cells are transfected with a pool of mutagenized plasmids and incubated at different temperatures. Many viral plaques can be harvested and screened for conditional lethal or host range phenotypes. The benefit of this approach is that one can select for a specific mutant phenotype such as temperature sensitivity, cold-sensitivity or altered host range. The disadvantages are the laborious and time-consuming screening tests and the loss of the mutant genotype in a cDNA molecule. Therefore, the mutation has to be mapped on the RNA level and reintroduced into the infectious cDNA to ensure that the introduced mutation is solely responsible for the observed phenotype (see below).

Currently, the primary nucleotide sequences and infectious cDNA copies of all three poliovirus serotypes, both neurovirulent and attenuated strains, have been reported (RACANIELLO and BALTIMORE 1981a, b; NOMOTO et al. 1982; OMATA et al. 1984; TOYODA et al. 1984; STANWAY et al. 1983, 1984). With the availability of these infectious cDNAs and the known nucleotide sequences, the poliovirus genome is available for genetic manipulations such that predetermined mutations can be introduced into the viral RNA. The resulting mutant phenotypes can then be examined for various aspects of viral growth.

2.2.2.1 Site-Directed Mutagenesis Methods

A variety of site-directed mutagenesis methods have been used to manipulate DNA molecules and the reader is referred to an excellent review (SHORTLE et al. 1981).

Most poliovirus mutants were obtained by manipulating the viral cDNA at existing restriction endonuclease cleavage sites. Restriction enzymes can be used which produce 5' protruding ends created by linearization of viral cDNA at any one of a large number of restriction sites. Nucleotide insertions can be obtained by repairing the protruding ends with the Klenow fragment of E. coli DNA polymerase (MANIATIS et al. 1982). Upon religation, a single codon insertion can be obtained if the protruding ends contained three nucleotides. Alternatively, synthetic DNA linker molecules can be cloned into various restriction enzyme sites in the cDNA; these synthetic DNA linkers can be designed to insert multiple codons in the correct translational frame. Small deletions can be created by trimming back 3' or 5' protruding ends with either T4 DNA polymerase or S1 nuclease, respectively (MANIATIS et al. 1982); larger deletions can be produced by trimming linearized cDNA with Bal31 exonuclease (MANIATIS et al. 1982).

In the absence of a convenient restriction endonuclease site, mutations can be introduced by oligodeoxynucleotide (oligo)-directed mutagenesis (SHORTLE et al. 1981). In this method, small mutations are created by annealing oligos bearing mismatches to gapped duplex (gd) DNA molecules (DALBADIE-McFARLAND et al. 1982). Following gap repair and transformation into E. coli, desired plasmids can be identified by colony hybridization using the oligos as a hybridization probe (WOOD et al. 1985). A more specialized case of site-directed mutagenesis is represented by the synthesis of pairs of complementary oligos bearing multiple mutations. These oligos can be annealed to each other and inserted into restriction sites in the viral cDNA (DEWALT and SEMLER 1987; KUHN et al. 1988a).

Recently, an amber codon-suppressing cell line has been generated in which a poliovirus amber mutant, created by site-directed mutagenesis, could be propagated (SEDIVY et al. 1987). The existence of this cell line opens the possibility of examining functional domains of the viral polyprotein by creating amber mutants which can be grown conditionally. Of course, this approach is somewhat limited by the extreme polarity of the poliovirus genome: any amber mutation prevents translation of every protein encoded 3' of the mutation.

2.2.2.2 Random Mutagenesis Methods

Random mutagenesis has been accomplished using two predominant experimental approaches. In the first method, double-stranded breaks are produced in circular DNA molecules by the use of pancreatic DNAse I in the presence of Mn^{2+}. These full-length linear molecules can then be manipulated to generate insertions by the addition of DNA linkers or deletions by exonuclease treatment (SHORTLE et al. 1981).

In the second approach, targeted regions of the genome are made single stranded either by forming a D-loop in the double-stranded plasmid (SHORTLE et al. 1981) or by producing gapped duplex molecules (KALDERON et al. 1982). The molecules can then be treated with sodium bisulfite, which deaminates cytosine residues, producing point mutations in the single-stranded region of the DNA molecules (PEDEN and NATHANS 1982; KALDERON et al. 1982).

Recently, KIRKEGAARD and NELSEN (1990a) have used a different method to generate small deletion mutants mapping in the viral capsid region. The protocol involves random nicking of the viral cDNA with pancreatic DNase I in the presence of ethidium bromide. The nicks are extended with T4 DNA polymerase in the presence of only three dNTPs. Gapped molecules are linearized with mung bean nuclease, recircularized, and transformed into *E. coli*. HeLa cells are then transfected with either a population of randomly deleted plasmids or with individual plasmids.

2.2.3 Genotypic Analysis of Mutants and Revertants

As described above, viral mutants can be obtained following transfection of individual mutated cDNA molecules or pools of mutated cDNAs. In the first case, the cDNA harboring the mutation is available for further analysis. Mutants arising from pooled transfection procedures, however, cannot be correlated to their parental cDNA. In either case, it is important to correlate the observed mutant phenotype with the introduced mutation.

If a cDNA copy of a mutant is available, the DNA sequences comprising the mutation must be re-engineered into an unmutagenized cDNA, followed by sequencing. This "mix and match experiment" is important because additional nucleotide changes could have occurred in the mutated cDNA during the mutagenesis procedures or during amplification in *E. coli*. In addition, the mutant virion RNA can be sequenced at the site of mutation to ensure that the lesion is present.

If a cDNA copy is not available due to the selection procedure used to obtain a mutant or revertant, the expected lesion is first roughly mapped by recombination tests or directly mapped by sequencing. Next, the lesion must be introduced into an unmutagenized cDNA, either by inserting a newly constructed cDNA copy spanning the expected lesion or by introducing the expected mutation via oligo-directed mutagenesis. The same methods can be applied to map second site revertants. The cDNA harboring the expected mutant genotype can then be transfected into appropriate host cells and scored for the mutant phenotype.

These tests, although cumbersome, are essential to produce, maintain, and study stable, well-defined viral mutants or revertants. Fortunately, these tests are now technically possible due to the existence of the infectious cDNA.

3 Structure and Function of the Poliovirus Genome

Sequencing of the entire poliovirus RNA (KITAMURA et al. 1981; RACANIELLO and BALTIMORE 1981a) and extensive biochemical analyses of viral polypepetides (reviewed in KITAMURA et al. 1981) have revealed the primary structure of the genome and its gene products (Fig. 2). The viral RNA is 7440 nucleotides (nt) in length and has a positive-strand polarity. A viral protein VPg (virion protein, genome linked) is covalently attached to the 5' end of the RNA. The 5' noncoding region contains 742 nt upstream of the AUG codon used for the initiation of translation in vivo. Translation produces a single polyprotein with a molecular weight of 247 000. After the large open reading frame is a short 3' noncoding region of 70 nt followed by a polyadenosine tail of varying length.

The functions of individual viral gene products have been studied by purification and enzymatic assays in vitro. In this way, structural proteins were identified by analysis of purified virions, and nonstructural proteins were purified which contained RNA polymerase and proteinase activities (RUECKERT 1985). However, the roles of many viral proteins and their precursor products remain unknown. Also, the functional role of the viral noncoding regions in translation and RNA amplification remained enigmatic until recently.

The aim of this section is to summarize results obtained from the study of well-defined mutants and to discuss what genetic approaches have taught us about the function of the viral genome.

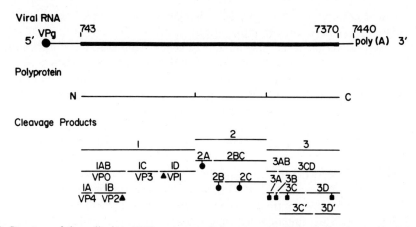

Fig. 2. Structure of the poliovirus RNA genome and cleavage map of the viral polyprotein. The structure of the genomic RNA is illustrated on the *top line*. The 5′ noncoding (nucleotides 1 to 743), the coding (nucleotides 743 to 7370) and 3′ noncoding (nucleotides 7370 to 7440) regions are indicated. The polyprotein and its cleavage products are shown using the L434 nomenclature system proposed for all picornaviruses (RUECKERT and WIMMER 1984). Mutations in codons leading to viable mutant phenotypes are indicated

3.1 The 5′ and 3′ Noncoding Regions

3.1.1 Features of the 5′ Noncoding Region

Several functions have been attributed to this region. It contains signals that are important for (a) the initiation of translation by a cap-independent mechanism (see the chapter by Sonenberg, this volume), (b) the initiation of replication for positive-strand RNA molecules (see the chapter by RICHARDS and EHRENFELD, this volume) and (c) the determination of the neurovirulent phenotype of poliovirus.

Sequence data have revealed three interesting features of this region. First, it contains an unusually long, 742 nt sequence that is highly conserved among the three poliovirus serotypes (TOYODA et al. 1984). Second, several apparently unused AUG codons (eight in Mahoney type 1) precede the translation initiation codon at position 743 (KITAMURA et al. 1981; RACANIELLO and BALTIMORE 11981a; PELLETIER et al. 1988). Third, a stable hairpin structure ($\Delta G^0 = -21$ kcal) close to the 5′ terminus of the positive strand has been predicted in the RNA of all three serotypes (LARSEN et al. 1981; CURREY et al. 1986). In addition, as many as 22 other RNA hairpin structures have been predicted in this region on the basis of computer-assisted analysis (RIVERA et al. 1988) and of biochemical tests (SKINNER et al. 1989).

The first genetic evidence suggesting the importance of the 5′ noncoding region in the viral life cycle was reported by RACANIELLO and BALTIMORE (1981b). They observed that a poliovirus cDNA molecule was not infectious when the first 115 basepairs (bp) were missing. Furthermore, inversion of the nucleotide sequences from nt 220 to nt 670 abrogated infectivity of the cDNA (V.R. RACANIELLO, personal communication).

Over the past several years numerous genetic studies have examined the role of the 5′ noncoding region. Two experimental approaches have been used: SEMLER and colleagues have analyzed this region with the use of chimeric enteroviruses, and, subsequently, several other groups have explored the function of this region by generating well-defined mutants.

3.1.1.1 The 5′ Noncoding Regions of Two Enteroviruses Are Exchangeable

SEMLER and co-workers noted a high degree of sequence homology (70%) between the 5′ noncoding (5NC) regions of poliovirus type 1 and coxsackievirus B3, another member of the genus *Enterovirus*. To test whether these regions can replace each other functionally, they substituted the poliovirus 5NC sequences from nt 220–627 with the 5NC sequences from nt 220–625 of coxsackievirus B3 (SEMLER et al. 1986). Upon transfection into monkey COS-1 cells, the chimeric cDNA yielded infectious virus that was temperature sensitive (ts). The recombinant virus produced 20-fold less viral RNA than wild-type poliovirus type 1 at the nonpermissive condition, 37° C. In addition, this mutant showed delayed kinetics in protein synthesis and inhibition of host cell protein synthesis at the nonpermissive temperature in comparison with wild-type virus. Although these pleiotropic effects make an assignment of any particular function to nt 220–627 difficult, this study revealed functional similarity between the 5NC regions of poliovirus type 1 and coxsackievirus B3.

To define the nucleotide sequences and structures responsible for the ts phenotype of the recombinant virus, JOHNSON and SEMLER (1988) isolated three independent revertants which produced viral proteins, viral RNA, and virus yields comparable to wild-type virus at 37° C. The 5NC of the revertants was sequenced and a four nucleotide deletion between nt 231–234, part of the region derived from coxsackievirus B3, was detected in each revertant. Assuming that this was the only lesion responsible for the revertant phenotype, this result suggests that sequences around nt 220 in the recombinant 5NC are important for viral replication and translation.

The authors extended their studies and generated chimeric viruses by replacing the 5NC sequences from nt 66–627 and 1–627 of poliovirus type 1 with the sequences from nt 64–630 and 1–630, respectively, from coxsackievirus B3. These chimeras, unlike the previous one, were not temperature sensitive for growth at 37° C and produced viral proteins in amounts similar to wild-type virus. Thus, the 5NC of at least two enteroviruses are functionally very similar.

3.1.1.2 Multiple Structural Elements in the 5NC Affect Translation and Amplification of the Viral RNA

The first mutation in this region was obtained during the synthesis of a cDNA copy of the 5NC region, most likely due to an error by reverse transcriptase

(RACANIELLO and MERIAM 1986). Upon transfection of the cDNA, RACANIELLO and MERIAM recovered a virus that was temperature sensitive for viral growth at 38.5° C. Sequencing revealed a deletion of nt 10 which has been predicted to be involved in base pairing at the bottom of the hairpin structure near the 5′ end of the viral genome (see above). The mutant was actinomycin D sensitive at all temperatures, and both positive- and negative-strand syntheses were severely reduced at the nonpermissive temperature. Protein synthesis was slightly delayed when compared with wild-type, but appeared not to be temperature sensitive. A nontemperature-sensitive, actinomycin D-sensitive revertant was isolated and found to contain a G → U base change at nt 34 which could restore the missing base pair to the stem of the hairpin structure. Thus, this RNA structure involving sequences from nt 10 to 34 seems to be important for amplification of the viral RNA. The continued actinomycin D sensitivity of the revertant could be due to the failure to restore a predicted hairpin structure in the negative strand.

Most studies designed to elucidate the functions of specific sequence elements in the 5NC have included mutants obtained by site-directed mutagenesis techniques (described above). We would like to discuss their genotypes and phenotypes in the remainder of this section (see Fig. 3), starting with the mutations at the very 5′ end of the genome.

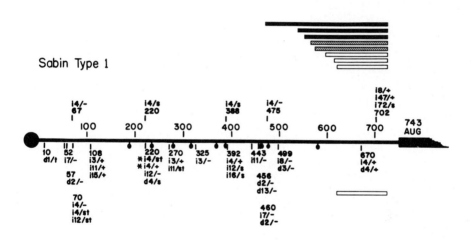

Fig. 3. Summary of mutations and mutant phenotypes mapping in the 5′ noncoding region of poliovirus type 1. Mutations and mutant phenotypes in the Sabin strain (above the viral genome) and in the Mahoney strain (below the viral genome) are shown. The description of the mutations and mutant phenotypes are as follows: nucleotide position of mutation, type of mutation (i, insertion; d, deletion), number of nucleotide changes, phenotype of mutant (+, wild-type; −, nonviable; t, temperature sensitive; s, small plaque); for example 10 d1/t represents a mutant bearing a single nucleotide deletion at position 10 displaying a temperature-sensitive phenotype. *Bars* indicate large deletion mutants with wild-type (*open*), small plaque (S *stippled*), and nonviable (*solid*) phenotypes. The positions of AUG codons are indicated (·). The asterisks (*) indicate the position of two mutants with the same genotypes but different phenotypes

Extensive mutagenesis on the 5NC of the Mahoney strain of poliovirus type 1 was performed by TRONO and co-workers (1988a) who accumulated a series of well-characterized mutants (Fig. 3). Their data show that different mutations created at the same site can result in multiple mutant phenotypes. For example, a deletion of four nt at position 67 results in a noninfectious cDNA, while an insertion of four nt at this position results in a mutant with a temperature-sensitive, small plaque phenotype (Fig. 3). This latter mutant has a primary defect in RNA synthesis, as shown by its ability to synthesize only 1% the amount of RNA as wild-type virus at both 32.5° C and 39.5° C. It was reported later that this mutant could support translation in vivo (TRONO et al. 1988b).

The sequences around nt 105 can tolerate insertions of 3, 11, or 15 nt; all these changes result in viruses with wild-type phenotypes (Fig. 3), suggesting that this region is not essential for viral replication and growth in tissue culture.

Perhaps the most interesting mutants have been obtained by manipulating sequences surrounding nt 220 (Fig. 3). TRONO et al. (1988a) generated a four nt insertion at nt 220, designated 5NC-13. This mutant displays temperature sensitivity at 39° C and has a small plaque phenotype on HeLa cells at the permissive temperature, 32°C. Synthesis of viral RNA, although somewhat lower than in a wild-type viral infection, was not temperature sensitive. However, protein synthesis and the inhibition of host cell translation was deficient at the nonpermissive temperature, suggesting that 5NC-13 has a primary defect in translation.

Interestingly, DILDINE and SEMLER (1989) reported the construction of the same four nt insertion mutant at nt 220 (220F) which displayed a ts-phenotype different from 5NC-13 (Fig. 3) in that 220F displayed a less than ten-fold higher titer at 33° C compared with 37° C. Several explanations to account for this apparent contradiction, such as undetected secondary lesions or the use of different host cells in the two cases, can be invoked.

Deletion of four nt at position 220 produced a temperature-sensitive mutant, 220D1 (DILDINE and SEMLER 1989; Fig. 3). Unlike 5NC-13, this mutant showed impairment of both RNA synthesis and protein synthesis at the nonpermissive temperature (37° C). A revertant (R2) of 220D1, selected for growth at 37° C, restored both RNA and protein synthesis at the nonpermissive temperature. Genotypic analysis of R2 revealed a 45 nt deletion between nt 184 and 228. The authors suggested that the 45 nt deletion in R2 possibly restored structural elements in the RNA which might be important for viral growth at the nonpermissive temperature.

Further support for the importance of structural elements surrounding nt 220 was provided by KUGE and NOMOTO (1987) who analyzed revertants of a ts-mutant mapping at nt 220 in the Sabin type 1 strain (Fig. 3). Sequencing of the 5NC of several revertants revealed the presence of several potential second site suppressor mutations. All the revertants had a U → C change at nt 186. In addition, three of the four revertants examined harbored a U → C change at nt 525, while the fourth had a C → A change at 480. However, it is not clear which of these changes is functionally responsible for the observed phenotypic suppression. Parenthetically, the guanosine at nt 480 is thought to be important for the attenuated phenotype of the Sabin type 1 virus (KAWAMURA et al. 1989).

The region from nt position 443 to 499 does not appear to be tolerant of mutations (Fig. 3), as shown by the inability of several groups to identify any viable mutants (DILDINE and SEMLER 1989; KUGE and NOMOTO 1987; TRONO et al. 1988a). This region appears to play a role in the attenuation phenotype of type 1 and type 3 polioviruses (EVANS et al. 1985; OMATA et al. 1986). Using a computer-assisted secondary structure analysis, EVANS and co-workers (1985) suggested that major alterations in RNA structures are caused by the $C \rightarrow U$ change at nt 472 in poliovirus type 3, which correlated well with the attenuation of neurovirulence in the Sabin strain. However, PILIPENKO et al. (1989) analyzed the structures of P3/Leon/37 and P3/Leon/12a$_1$b (wild-type and Sabin strains respectively) virion RNAs by enzymatic and chemical cleavage reactions and failed to observe any major structural differences between the RNAs. Interestingly, the $C \rightarrow U$ base change (nt 472) in type 3 poliovirus does affect the translation efficiency of the RNAs in vitro (SVITKIN et al. 1985, 1988) and the ability of chimeric viruses (type 3 5NC/type 2 coding region) to replicate in the brains of mice (LA MONICA et al. 1987). Recent findings by MINOR and DUNN (1988) point to a potential role of nucleotide sequences 471 to 490 in the ability of poliovirus to replicate in the gut.

A region with a high sequence variability among the three serotypes is located around nt 702 (TOYODA et al. 1984), suggesting that this region may not harbor sequence elements crucial for replication or translation. To test this, KUGE and NOMOTO (1987) inserted eight, 47 or 72 foreign nucleotides at nt 702 in the Sabin type 1 genome (Fig. 3). Only the 72 nt insertion mutant displayed a phenotype slightly different from wild type, suggesting that there are limits to the size of insertions (KUGE et al. 1989). Next, the authors constructed a series of deletion mutants (KUGE and NOMOTO 1987). Deletions from nt 726 to nt 600, 622 or 629 had no effect on the viability of the viruses. Deletions from nt 726 to nt 564 or 570 also produced viable virus, but these mutants displayed small plaque phenotypes. Deletions from nt 726 to nt 480, 534 or 551 did not produce viable virus. Based on these studies, it appears that sequences from nt 570 to 726 are not essential for growth of poliovirus in tissue culture.

The phenotypes and known genetic lesions of many mutants in the 5' noncoding region of Sabin and Mahoney type 1 polioviruses are compiled in Fig. 3. In summary, the characterization of many defined viral mutants have identified several genetic elements of the 5NC involved in viral replication, translation, and neurovirulence. The extreme 5' end of the viral RNA seems to be important in the amplification of viral RNA, while sequences 3' of nt 200 are necessary for efficient cap-independent translation. The latter idea has been further substantiated by the finding that nucleotide sequences from nt 186 to 680 can direct the internal binding of ribosomes to the viral RNA (PELLETIER and SONENBERG 1988; discussed in Chap. 2, this volume).

Sequence analysis of phenotypic revertants of the Sabin mutants at position 220 seem to indicate that certain mutations between nt 186 and 525 can affect the neurovirulent phenotype of poliovirus. However, mix and match experiments (see above) to ensure that a particular nucleotide alteration was responsible for the phenotypic reversion were not performed; thus, additional lesions in the remainder of the revertant genomes could contribute to the virulent phenotypes. The position of the second site mutations and the phenotypes of the attenuated viruses indicate that the attenuated genome's inability to grow in motor neurons may be due to defects in translation. This is supported by the finding that attenuated and virulent genomes differ in in vitro translation efficiencies (SVITKIN et al. 1985).

3.1.1.3 An In Vitro Biochemical Approach to Define the Function of the 5′ Noncoding Region

Genetic studies may suggest that higher order structures in the RNA are required for the proper functioning of the 5NC in replication, translation and neurovirulence. In vitro translation assays have aided in suggesting effects of mutations in this region on translation (SVITKIN et al. 1985; PELLETIER et al. 1988; TRONO et al. 1988b), however, direct biochemical evidence is still missing. We have used a molecular genetic approach to identify cellular and viral proteins that interact with this region.

   We have employed a gel retention assay to detect complexes of polypeptides with specific regions in the viral RNA. This technique was originally employed to detect protein-RNA interactions in ribosomes (HJERTEN et al. 1965; DAHLBERG et al. 1969). Recently, the gel retention assay has been refined by KONARSKA and SHARP (1986) to analyze RNA-protein interactions in splicing complexes.

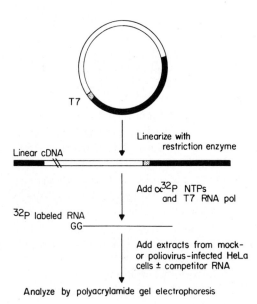

Fig. 4. RNA gel retention assay to study protein-RNA interactions. A plasmid containing viral cDNA (*solid bar*) downstream of the T7 φ10 promoter (*stippled bar*) is shown. The strategy for the RNA gel retention assay is discussed in the text

Depicted in Fig. 4, the assay is based on the observation that free nucleic acids migrate faster through a polyacrylamide gel matrix than nucleic acids complexed to proteins; thus, specific nucleic acid-protein complexes can be identified. Briefly, radioactively labeled viral RNA molecules are synthesized in vitro by T7 RNA polymerase and incubated with cytoplasmic extracts from infected or uninfected cells in the presence of a 250-fold molar excess of nonspecific competitor RNA (polyACU or *E. coli* ribosomal RNA). A ten-fold molar excess of a specific competitor RNA (the same RNA fragment, but unlabeled) is added in a separate reaction. After a short incubation at 30° C, the reaction products are analyzed by polyacrylamide gel electrophoresis. Figure 5 displays an autoradiograph of a gel retention assay, using a 42 nucleotide viral sequence (nt 178 to nt 220) as a radiolabeled template RNA. Lanes 1 and 2 show the migration of the labeled fragment in the presence of polyACU and polyACU + specific competitor, respectively. Upon incubation with polyACU and cytoplasmic extracts from wild-type infected (lane 3) or uninfected (lane 4) HeLa cells, slower migration of the labeled fragment is observed. However, addition of specific competitor RNA abolishes the "retention" (lanes 5 and 6) indicating that the complex formed is specific for that particular RNA. Subsequently, we have shown that the retardation is proteinase sensitive. Thus, a protein factor present in both uninfected and infected cell extracts, binds specifically to a 42 nucleotide sequence element in the 5NC (NAJITA and SARNOW, manuscript submitted). The binding site of this factor lies within a 45 nt region deleted in R2, a viable revertant of a mutant whose original lesion is at position 220 in the viral 5' NC (DILDINE and SEMLER 1989). Therefore, it is possible that binding of this factor is not

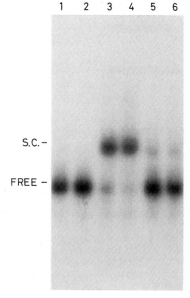

Fig. 5. Identification of a cellular factor interacting with poliovirus sequences nt 178–220. Radiolabeled RNA representing poliovirus sequences 178–220 was incubated with cytoplasmic extracts from poliovirus-infected (*lanes 3* and *5*) or uninfected (*lanes 4* and *6*) cells in the absence (*lanes 3* and *4*) or presence (*lanes 5* and *6*) of specific competitior RNA (see text). In a control experiment, radiolabeled RNA was incubated with *lane 2*) or without (*lane 1*) specific competitor RNA. Protein-RNA complexes are analyzed by polyacrylamide gel electrophoresis and an autoradiograph is shown. The migration of unbound RNA-protein complexes (*s.c.*) is indicated

necessary to support viral growth in tissue culture; alternatively, R2 may have regenerated structural elements in the RNA which mediate the binding of the factor. The latter would suggest that factor binding is not solely dependent on primary nucleotide sequences.

Thus, a gel retention assay can be used to investigate the binding of cellular and viral factors to various regions of the 7500 nucleotide viral RNA. The minimal binding sequences can be mapped by using competitor RNA molecules of different lengths. In addition, potential structural elements in the RNA which may be important for protein binding can be tested using RNA template molecules modified by site-directed mutagenesis methods or by base-specific chemical probes.

3.1.2 Genetics of the 3′ Noncoding Region

The 3′ noncoding region (3NC) consists of approximately 70 nt downstream of the large open reading frame, and is followed by a polyadenosine tail. A role for this region in directing the synthesis of negative-strand RNA has been suggested by the phenotypes of mutants bearing nucleotide insertions at nt 7387 (SARNOW et al. 1986). Insertions of two or ten nucleotides at this position gave rise to mutants displaying a phenotype similar to wild-type. In contrast, an eight nucleotide insertion gave rise to a ts mutant, 3NC202, that did not produce detectable amounts of negative-strand RNA at the nonpermissive temperature (39.5° C). Curiously, the nucleotide insertions map to a region where IIZUKA et al. (1987) have predicted the formation of an RNA pseudoknot. According to their computer-assisted structural model, the insertions would lengthen the hairpin

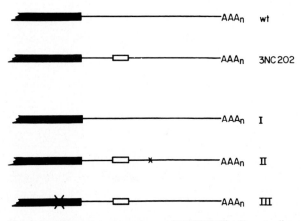

Fig. 6. Possible revertant genotypes of 3NC202. The 3′ noncoding regions of wild-type and mutant 3NC202 are indicated at the *top* of the figure. The *open box* represents the eight nucleotide insertion in 3NC202. Three possible genotypes of 3NC202 revertants are indicated: (*I*) Elimination of insertion (true revertant), (*II*) second site suppressor mutation in noncoding region, and (*III*) second site suppressor mutation in coding region

loop of the pseudoknot. Pseudoknot RNA structures have been observed near the 3' termini of most plant viruses that have single-stranded RNA genomes (PLEIJ et al. 1987). Although their physiological function is not clear, it has been speculated that pseudoknots may have a role in the amplification of the viral genomes (PLEIJ et al. 1987). In addition, it has been suggested that tRNA-like structures may have played a role in the early evolution of self-replicating RNA genomes (WEINER and MAIZELS 1987).

To understand the structural and functional requirements of this region in the initiation of negative-strand RNA synthesis, we have isolated and characterized revertants of 3NC202. Three possible revertant genotypes are diagrammed in Fig. 6. A simple back-revertant would eliminate the insertion (I); a second site suppressor mutation in the 3NC would reveal information about structural requirements for initiation of negative-strand synthesis (II), and a second site suppressor mutation in the coding region might identify viral polypeptides that interact with this region (III). Thus far, we have isolated revertants of class II (revertants R3 and R7) and class III (R6). R3 and R7 are being used to examine the potential role of the proposed pseudoknot structure. R6 has a second site suppressor mutation in P3 and is currently being sequenced and recloned.

3.1.3 The Polyadenosine Tail

The virion RNA contains a polyadenosine (poly (A)) tail at its 3' end with an average length of 75 nt (ARMSTRONG et al. 1972; YOGO and WIMMER 1972). Upon infection, cytoplasmic viral RNA can be detected containing 150 to 200 adenosine residues (SPECTOR and BALTIMORE 1975a), suggesting the presence of a poly(A) polymerase activity in the cytoplasm of infected cells that is active on poliovirus RNA (SPECTOR and BALTIMORE 1975b).

The role of these 3' end sequences in the infectivity of poliovirus was examined by transfection of manipulated viral RNA molecules into mammalian cells. SPECTOR and BALTIMORE (1974) have reported that deadenylated virion RNA molecules containing 20 or fewer adenosine residues is only 5% as infectious as intact virion RNA. Recently, the infectivity of RNA synthesized in vitro by T7 RNA polymerase bearing different 3' end sequences was examined in detail (SARNOW 1989). It was found that RNA molecules containing 12 adenosine residues and 17 additional cytosine residues at their 3' ends were 2% as infectious as virion RNA, whereas RNA molecules with only four additional residues, CGCG, after the adenosine tail showed an increased activity to 10% of virion RNA. If the latter RNA molecules were polyadenylated in vitro, specific infectivities comparable to that of virion RNA were obtained. These data clearly demonstrate an effect of the length of the poly(A) tail on the infectivity of virion RNA. However, it is not clear what the primary role of these terminal sequences are. They may be important in the initiation of either negative-strand RNA synthesis (see the chapter by RICHARDS and EHRENFELD, this volume) or RNA stability in the infected host cell. These possibilities are not mutually exclusive and further studies are clearly needed.

3.2 The P1 Region

The P1 region, the 5′ terminal segment of the viral coding region (Fig. 2), encodes the structural proteins VP4, VP2, VP3, and VP1 in that order. Sixty copies of each individual capsid protein are found in the final infectious virion (reviewed by PUTNAK and PHILLIPS 1981). During the last steps of morphogenesis, VP4 and VP2 are generated by proteolytic cleavage of VP0, their precursor polypeptide. It has been proposed that this cleavage may occur by an autocatalytic process in which the viral RNA participates (ARNOLD et al. 1987).

The elucidation of the crystal structure of poliovirus (HOGLE et al. 1985) and the availability of an infectious cDNA clone have helped in the understanding of the molecular basis of poliovirus antigenicity. In this way, intertypic recombinant viruses involving antigenic sites and neutralization escape mutants have been isolated and mapped (reviewed in Chapter 5 by MINOR, this volume).

Several defined capsid mutants have been obtained by manipulating infectious cDNA molecules. In one study, synthetic oligodeoxynucleotides were used to insert three or six codons into the coding region of VP1 (COLBERE-GARAPIN et al. 1988). The site of insertion encodes amino acid 100 in VP1, which is contained within an external polypeptide loop structure on the virion, involved in a characterized antigenic site (HOGLE et al. 1985). The mutants were viable and, interestingly created a new antigenic site against which a new class of neutralizing antibodies could be raised. This finding opens the possibility of constructing virions containing antigenic sites with a particular neutralization specificity.

Using a random mutagenesis method, KIRKEGAARD and colleagues (1990a,b) have manipulated infectious poliovirus type 1 cDNA to generate small deletions in the viral genome. Two host range and one ts mutant have been isolated that contain lesions in P1; all three mutants have been reconstructed to confirm that these lesions are responsible for the mutant phenotypes.

One mutant, VP1-101, bears a deletion of amino acids 8 and 9 in VP1. This mutant displayed a plaque-reduction phenotype on monkey CV-1 cells relative to human HeLa cells. Its phenotype could be traced to a decreased ability to enter CV-1 cells. A second mutant, VP1-102 bears a deletion of amino acids 1 to 4 in VP1, and displayed CV-1 cell-specific defects in both cell entry and virion morphogenesis. This indicates that the extreme amino terminus of VP1, unresolved in the crystal structure of the poliovirion (HOGLE et al. 1985) is involved in both cell entry and viral assembly (KIRKEGAARD 1990b).

A ts-mutant in P1, VP2-103, bears a single nucleotide change in the coding sequence of VP2, altering amino acid 75. RNA and protein synthesis were similar to wild-type at the nonpermissive temperature. However, no cleavage of VP0 to yield VP4 and VP2 was observed at 39.5°C and provirion-like particles accumulated (COMPTON et al., manuscript submitted). Therefore, VP2-103 appears to be a maturation mutant, defective in VP0 processing, and can perhaps be used to test the involvement of the virion RNA in VP0 proteolysis (see above).

Construction of more defined capsid mutants and subsequent comparison of their phenotypes with the known structure of the virion will undoubtedly shed further light on the function of the viral capsids. Dynamic processes such as viral uncoating, packaging, and neutralization can be best studied by a combination of genetic and structural analysis.

3.3 The P2 Region

The P2 region, the middle segment in the viral coding region, encodes several nonstructural polypeptides: 2A, 2BC, 2B, and 2C. Biochemical and genetic experiments have implicated each of these processed proteins in viral replication. As discussed below, we are now equipped with a variety of well-defined mutants that will allow us to study the functions of these proteins in more detail.

3.3.1 Proteinase 2A

Biochemical experiments employing the in vitro translation of RNA molecules encoding 2A provided the first evidence that 2A is a proteinase that cleaves the viral polyprotein between certain tyr/gly amino acid residues (TOYODA et al. 1986). Subsequently, this was confirmed by co-purification of 2A with proteinase and esterase activities (KOENIG and ROSENWIRTH 1988). One 2A-specific cleavage occurs intramolecularly, removing P1 from the P2/P3 region. The second 2A-specific cleavage appears to be intermolecular, cleaving 3CD into 3C' and 3D' (see discussion by WIMMER et al. 1987). The P1/P2 cleavage liberates capsid precursors from the rest of the polyprotein, a prerequisite for the second viral proteinase, 3C (see the Chapter 3 by LAWSON and SEMLER , this volume) to cleave the P1 precursor (NICKLIN et al. 1987; YPMA-WONG et al. 1988a). Site-directed mutagenesis of the tyr/gly recognition site has been shown to prevent the cleavage of 3CD to 3C' and 3D', but this had no obvious effect on virus growth in tissue culture (LEE and WIMMER 1988).

To study the function of 2A in vivo, BERNSTEIN et al. (1985) constructed a mutant, 2A-1, that contained a single amino acid insertion in the N-terminal region of 2A corresponding to nt position 3690 and displayed a small plaque phenotype. 2A-1 was unable to selectively inhibit host protein synthesis, a process normally accomplished by wild-type virus within 2 to 3 hours postinfection. However, 2A-1 was apparently able to process the P1/P2 cleavage site. Thus, the defective function in 2A-1 is not due to the inhibition of 2A proteinase activity at the P1/P2 cleavage site. Viral protein synthesis in 2A-1 infected human HeLa cells was depressed, possibly due to its inability to effectively compete with cellular mRNAs for the translational machinery. Curiously, 2A-1-infected cells display a global (viral and cellular) inhibition of protein synthesis in HeLa cells at 4 h postinfection that is qualitatively similar to that seen late in wild-type virus infection. This indicated that specific (cellular) and nonspecific (cellular and viral) inhibition of protein synthesis occur by different mechanisms. Complementation studies with several defined mutants showed that the 2A-1 defect could be effectively rescued (see Sect. 5.2); therefore, only small amounts of the translation inhibitor must be required for specific inhibition of cellular translation by poliovirus, a process in which 2A is directly implicated.

Bernstein and co-workers examined why cellular protein synthesis was not inhibited early in 2A-1 infected cells. To this end, they analyzed the fate of p220, a protein of the cap-binding complex. P220 is cleaved in wild-type infected cells (ETCHISON et al. 1982, 1984); however, negligible levels of p220 cleavage were detected in 2A-1 infected cells. This supports the hypothesis that p220 cleavage is an important part of poliovirus-mediated inhibition of host translation. Several investigators have shown subsequently that intact 2A, biochemically defined by its ability to cleave the P1/P2 junction in an in vitro translation system, is required for p220 cleavage, most likely indirectly, through the activation of a cellular proteinase (KRAUSSLICH et al. 1987; LLOYD et al. 1988).

3.3.2 Functions of 2BC and Its Cleavage Products 2B and 2C

Very little is known about the function of polypeptide 2BC and its cleavage products 2B and 2C. 2BC has been found associated with membrane-bound replication complexes in infected cells (BIENZ et al. 1983, 1987, 1990). It has been proposed that this polypeptide may somehow function in the induction of unique membranous structures which support viral replication (BIENZ et al. 1983).

The only genetic evidence to date suggesting that 2B or 2BC may have a role in viral replication are the phenotypes of two mutants, 2B-201 (BERNSTEIN et al. 1986) and 2B-11 (LI and BALTIMORE 1988). Mutant 2B-201 bears a two amino acid insertion in the 2B coding region (at nucleotide position 3913) resulting in a severe defect in RNA synthesis (BERNSTEIN et al. 1986). This mutant synthesized ten fold less RNA than wild-type virus at all temperatures tested; this defect was accompanied by diminished viral protein production as well as a ten fold decrease in virus yield. As a result, 2B-201 displayed a small plaque phenotype on HeLa and CV-1 cells at all temperatures (K. JOHNSON and P. SARNOW). Interestingly, the defect in 2B-201 could not be rescued by any other mutant tested. This lack of complementation could be explained by a sequestering of 2B (or 2BC) through immediate association with other proteins or with the viral RNA from which it was translated. An alternative explanation is that the insertion mutation caused an aberrant polypeptide folding. This is unlikely because the polyprotein processing pattern in mutant infected cells was apparently normal. Of course, all mutations in the poliovirus genome could potentially effect either the RNA structure and sequence, or the proteins encoded there.

A second mutant, 2B-11, has been characterized that bears a two amino acid insertion in the 2B coding region corresponding to nt position 3933 (LI and BALTIMORE 1988). 2B-11 also exhibited a noncomplementable defect in RNA synthesis.

Our understanding of the function of 2C in replication has been shaped largely by analyzing poliovirus mutants selected for growth in the presence of guanidine hydrochloride. Replication of wild-type virus is selectively inhibited by guanidine concentrations as low as 0.1 mM apparently by blocking the synthesis of single-stranded RNA and replicative intermediate RNA (CALIGUIRI and TAMM 1968; NOBLE and LEVINTOW 1970). It is not clear at what stage guanidine exerts its effect. However, there is evidence that the initiation step of RNA synthesis (CALIGUIRI and TAMM 1968; TERSHAK 1982) or possibly the release of newly completed RNA molecules (HUANG and BALTIMORE 1970) may be blocked. Guanidine-resistant (g^r) mutants spontaneously arise in wild-type virus-infected cells grown in the presence of guanidine. Recombinational mapping of several such mutants showed that guanidine resistance was a viral function mapping to the 3′ terminal region of the genome (EMINI et al. 1984). The guanidine locus in poliovirus was further biochemically mapped to the 2C coding region through electrofocusing and peptide mapping of viral polypeptides produced in g^r mutant-infected cells (ANDERSON-SILLMAN et al. 1984).

In an effort to understand the genetic basis of the 2C locus with the g^r phenotype, PINCUS et al. (1986) sequenced six individual isolates resistant to 2mM guanidine. All six mutants contained an amino acid substitution at position 179 in the 2C protein. They were able to show by mix and match experiments (see Sect. 2.2.3) that the single amino acid change at 179 was sufficient to confer the g^r phenotype (PINCUS and WIMMER 1986).

Passage of g^r viruses that were resistant to either a high (2 mM) or a low concentration (0.5mM) of guanidine yielded mutants that were dependent on the drug for growth. In each case, the guanidine-dependent mutant (g^d) contained the parental amino acid substitution plus one additional amino acid change within the 2C coding region (PINCUS et al. 1987b). Interestingly, all of the g^r and g^d mutants sequenced displayed mutations which were contained within an 85 amino acid stretch of the 329 residue 2C polypeptide. This region exhibits a striking sequence homology among several picornaviruses (ARGOS et al. 1984). The effect of guanidine on viral RNA amplification and the tight association of 2C with 3AB (see Sect. 3.4.1) in membrane-bound replication complexes, support the hypothesis that 2C may be an essential component of viral replication (BUTTERWORTH et al. 1976; SEMLER et al. 1982; TAKEGAMI et al. 1983b; PINCUS et al. 1987a).

To further investigate the role of 2C in the virus life cycle, LI and BALTIMORE (1988) have constructed defined mutants throughout the 2C coding region employing linker insertion mutagenesis. Of ten mutants constructed, only two were viable. Five of ten mutants mapped to the putative guanidine locus and were nonviable. 2C-31 contains a four amino acid insertion (at nucleotide position 4886) and exhibited a small plaque, ts phenotype on HeLa cells. 2C-32 contains a two amino acid insertion (at nucleotide position 4908) and formed small plaques on HeLa cells at 32.5° C and 39° C. Both mutants exhibited reduced RNA synthesis, protein production, and virus yield at 32.5° C and 39.5° C compared to wild-type virus. Despite the low amount of viral protein in 2C-31 infected cells at 39° C, the P2 region was properly processed.

Subsequently, LI and BALTIMORE (1988) analyzed the properties of 2C-31 infected cells in detail and made two interesting observations. First, temperature shift studies showed that upon shifting to the nonpermissive temperature, RNA synthesis was essentially stopped within 30 min. This result indicates that 2C or its precursor 2BC is continually required for RNA synthesis, although it is not clear whether altered protein conformation or diminished protein production is responsible for this effect. Second, no accumulation of double-stranded RNA molecules was noticed, an observation found in cells infected with wild-type virus after the addition of guanidine (BALTIMORE 1968). Thus, the authors propose that 2C may perform more than one function. Interestingly, LI and BALTIMORE have recently reported a revertant of 2C-31 (designated 2C-31R1), containing the parental linker insertion plus two additional point mutations within the 2C coding region, that possesses a cold sensitive defect for virion uncoating (LI and BALTIMORE 1990).

3.4 The P3 Region

The P3 region encompasses the carboxy-terminal segment of the poliovirus coding region. Gene products of P3 are posttranslationally cleaved into seven polypeptides: 3AB, 3A, 3B, 3C, 3D, 3C', and 3D'. Polypeptides 3B, 3C, and 3D have been most thoroughly studied both biochemically and genetically, due to their rather prominent roles in viral replication.

3.4.1 Functions of 3AB and Its Cleavage Products 3A and 3B

Several unique properties of the polypeptide 3AB have generated speculation about its role in vial RNA amplification. First, protein 3AB is the precursor of 3B (VPg), the 22 amino acid protein covalently linked to the 5' termini of positive- and negative-stranded viral RNAs (NOMOTO et al. 1977b; PETTERRSON et al. 1978). Second, 3AB has been found tightly associated with membrane-associated replication complexes in infected cells (TAKEGAMI et al. 1983b), probably by virtue of a N-terminal stretch of hydrophobic amino acid residues (SEMLER et al. 1982; TAKEGAMI et al. 1983b).

Recently, genetic evidence has shown a functional role of 3AB in viral replication. BERNSTEIN and BALTIMORE (1988) reported the construction of a viable mutant (designated 3A-2) bearing a three amino acid insertion in the 3A coding region (at nucleotide position 5152). This mutant formed small plaques on monkey Vero cells at $32.5°$ C but wild-type size plaques at $39.5°$ C. Thus, 3A-2 displayed a cold-sensitive phenotype. The mutation caused a marked decrease in the synthesis of positive- and negative-sense RNA at the nonpermissive temperature. Two interesting observations emerged from temperature shift and complementation studies: (a) When the temperature was shifted from $39.5°$ C to $32.5°$ C at 3 h post infection, RNA synthesis was not impeded. Thus, 3AB is either required primarily during the early, exponential phase of RNA synthesis, or very little of this protein is required during the late phase; (b) the defect in 3A-2 could be complemented by the replication-deficient mutants 2B-201 and 3D-3ts. Moreover, the ability of these mutants to rescue 3A-2 was proportional to the amount of protein produced, suggesting that the function that is defective in 3A-2 is overproduced during wild-type virus replication, and that only small quantities are needed to complement the 3A-2 defect (BERNSTEIN et al. 1986).

Several roles have been proposed for 3B (VPg) in the virus life cycle:

1. VPg may act as a primer for viral RNA (see Chapter 4 by RICHARDS and EHRENFELD this volume). According to this model, uridylylation of 3AB at the single tyr residue in the 3B segment would enable it to act as a primer for RNA synthesis. Subsequent cleavage of VPg from the precursor 3AB would yield the covalent RNA-protein product observed in infected cells, (NOMOTO et al. 1977a; TAKEGAMI et al. 1982a, b). According to this mechanism it is possible that the distance between a prominent hydrophobic sequence in 3AB and the VPg sequences at the C-terminus may be important for positioning VPg as a primer. To test this, KUHN et al. (1988b) constructed three mutants, each bearing a five amino acid insertion just upstream of the 3A/3B cleavage site. Unfortunately, no viable viruses were recovered.

2. VPg may possess a nuclease activity. An RNA hairpin-priming model of initiation of viral RNA synthesis (see Chapter 4 by RICHARDS and EHRENFELD, this volume) predicts that 3B should participate in a nucleolytic attack mechanism such that 3B would become covalently attached to the 5′ end of the nascent minus strand concomitant with cleavage of the RNA hairpin. Indeed, TOBIN et al. (1989) provided evidence that 3B attachment to poliovirus RNA can be mediated in vitro through a self-catalyzed transesterification reaction, similar to that observed with self-splicing RNA molecules (CECH and BASS, 1986). Defined mutants in 3B will help to test this model.

3. VPg may play a role in encapsidation. It has been demonstrated that VPg is found only on newly synthesized RNA molecules, and is not found on positive-sense RNA associated with ribosomes (NOMOTO et al. 1977a). In addition, only newly synthesized RNA is incorporated into virions (CALIGUIRI and TAMM 1968), hinting at a role of VPg in packaging of the viral genome. This hypothesis could be substantiated by the discovery of packaging mutants mapping in 3B.

In an attempt to study the function of 3B, KUHN et al. (1988a) used a novel mutagenesis approach to construct mutations in the 3B coding region. These investigators created several new restriction enzyme sites within and adjacent to the 3B coding region. These new sites enabled them to insert a number of synthetic double-stranded oligonucleotides bearing specific mutations in the 3AB coding region. In this way, a small region of the genome can be mutagenically saturated in a systematic manner. A large collection of mutations in 3AB and 3B revealed that: (a) the position of the tyrosine appears to be essential for its function; (b) substitution of a tyrosine for threonine at position 4 yielded nonviable virus, but in this case the polyprotein folding was apparently altered such that several 3C mediated cleavages were not made (KUHN et al. 1988b); (c) the arginine at position 17 is essential for virus viability, however, many other mutations can be tolerated in 3B.

Unfortunately, no conditional mutants were obtained from this study, which would have provided tools to study directly the function of 3B. Continued genetic analysis of the 3AB coding region is necessary to further understand the role of this region in various processes in infected cells.

3.4.2 Functions of 3CD and Its Cleavage Products 3C, 3D, 3C', and 3D'

Polypeptide 3CD is processed at two alternate cleavage sites: cleavage at a gln/gly peptide bond yields 3C and 3D; alternatively, cleavage between tyr/gly produces 3C' and 3D'. Biochemical evidence has suggested that 3CD and 3C are proteinases which cleave the polyprotein at gln/gly pairs in the P1 capsid precursor (YPMA-WONG et al. 1988b) and P2–P3 precursors, respectively (HANECAK et al. 1982). 3D is the viral RNA-dependent RNA polymerase (VAN DYKE and FLANEGAN 1980) which copies the viral genome. The functions of 3C' and 3D' are not known.

3.4.2.1 Proteinases 3CD and 3C

Proteins 3CD and 3C are the viral proteinases that cleave the polyprotein at selected gln/gly peptide bonds and are responsible for most of the cleavages of the polyprotein. The analysis of mutants in in vitro translation systems has been extremely useful in defining some aspects of substrate recognition by these proteinases. These experiments are discussed in Chapter 3 by LAWSON and SEMLER (this volume).

DEWALT and SEMLER (1987) were the first to report the construction of viable mutants in the 3C coding region. They used oligonucleotide-directed mutagenesis to introduce multiple amino acid substitutions at three different codons in 3C. Six viable mutants were isolated, all of which displayed small plaque phenotypes on HeLa cells (discussed in the chapter by LAWSON and SEMLER, this volume).

In one mutant, Sel-3C-02, amino acid 54 in 3C is changed from a valine to an alanine residue. Low levels of 3C-proteinase and 3D-RNA polymerase were produced in mutant-infected cells, while the levels of 3C' and 3D' were as high as in wild-type infected cells. In addition, the virus yield in a single infectious cycle was nearly the same as in wild-type infected cells. The authors suggest that low levels of 3D are sufficient for production of high virus yields. This speculation has been made in studies of mutants which show a decreased level of translation (BERNSTEIN et al. 1985; TRONO et al. 1988a).

3.4.2.2 RNA-Dependent RNA Polymerase 3D

Protein 3D is the viral RNA-dependent RNA polymerase that copies the viral genome. The biochemical properties of 3D are discussed in the chapter by RICHARDS and EHRENFELD (this volume). It has been puzzling that 3D apparently copies only the viral RNA in infected cells, but several nonviral polyadenylated

RNA templates can be copied in vitro (see the chapter by RICHARDS and EHRENFELD, this volume). Thus, the specificity of RNA synthesis in vivo has yet to be mimicked in vitro; it is possible that the viral genome is replicated by a *cis*-acting replication complex in infected cells.

To date, only two viable well-defined mutants have been constructed in 3D; both have conditional phenotypes. BERNSTEIN and co-workers constructed a mutant, 3D-3, that contains a single amino acid insertion (corresponding to nucleotide position 7048) and exhibited a small plaque phenotype on CV-1 cells at all temperatures (BERNSTEIN et al. 1986). 3D-3 synthesized a nearly wild-type amount of RNA, but showed a slight time delay in RNA accumulation. Protein synthesis, polyprotein processing, and virus yield were comparable to wild-type. A ts variant of 3D-3, designated 3D-3ts, which has an uncharacterized additional mutation, has growth properties similar to 3D-3 but yields fewer infectious particles at 39.5°C. Complementation tests with other mutants showed that the 3D function could not be supplied in *trans* to these 3D mutants. A possible explanation for this observation is that the viral polymerase may be able to interact only with its own template; upon cleavage from 3CD, it may immediately associate with other proteins and be recruited directly into the replication complex.

AGUT and colleagues (1981) isolated a ts mutant, ts035, bearing two point mutations in 3D. Upon reconstruction of the mutant genotype into an infectious cDNA clone, it was found that only one of the mutations, replacing asparagine 426 with aspartate, was responsible for the ts phenotype (AGUT et al. 1989). This mutant is severely defective in RNA synthesis and virion progeny formation at the nonpermissive temperature; ts305 could be complemented by wild-type virus and mutant ts221, whose RNA synthesis is also impaired at the nonpermissive condition (AGUT et al. 1981). However, ts305 could not be complemented by another ts mutant, ts203. It should be noted that ts221 and ts203 are not well-defined mutants and most likely bear multiple lesions. In addition, this study did not examine the possibility that recombinants may have arisen during the complementation assay.

In any case, to answer whether the viral RNA polymerase acts in *cis* or in *trans* requires further genetic analysis. To address this question, an in vivo genetic approach is discussed in Sect. 5.

3.4.2.3 The Alternate Cleavage Products of 3CD: 3C' and 3D'

The viral proteinase 2A cleaves 3CD at a tyr/gly pair (amino acids 147/148 in 3D) to yield 3C' and 3D'. The observations that different strains of poliovirus produce different amounts of 3C' and 3D' in infected cells, and that the Sabin type 3 strain lacks this alternate cleavage site entirely, lead to the notion that 3C' and 3D' may not be essential for viral growth.

To test this, LEE and WIMMER (1988) changed the 3C'/3D' cleavage site in poliovirus type 1 by site-directed mutagenesis. Substitution of tyrosine-148 by phenylalanine produced a precursor that could be cleaved by the 2A proteinase, while substitution of threonine-147 by a alanine produced a precursor that could not be cleaved. However, both mutations produced infectious viruses with phenotypes similar to wild type. Therefore, 3C' and 3D' do not play major roles in viral growth, at least in tissue culture.

4 Genetic Recombination in Poliovirus

The outcome of a genetic recombination event is a progeny genome with demonstrable contributions from both parents. Genetic recombination presumably results from a physical interaction between different parental genomes in mixedly infected cells; the nature of this interaction appears to differ between RNA and DNA genomes.

Historically, genetic recombination tests have been used to establish proximity and linkage between different loci to construct genetic maps of genomes.

4.1 The Recombination Test

A recombination test is initiated by infecting cells both singly or mixedly with two conditional-lethal mutants under permissive conditions. After one replication cycle, the viral yields in the singly (for both mutants) and mixedly infected cells are determined under permissive and nonpermissive conditons. The virus yields from singly infected cells under nonpermissive conditions are used to estimate the amount of leakage and reversion of the parental mutants. A higher virus yield from mixedly infected cells indicates that genetic recombination has occurred, if the resulting viruses display a recombinant phenotype. The recombination frequency is the yield of recombinants in the mixed-infection minus the sum of the yields of the single infections under nonpermissive conditions, divided by the total yield of the mixed infection, assayed under permissive conditions. It is important to test that apparently recombinant plaques are due to recombinant phenotypes and are not formed as a result of complementation in the mixed infection. This can be easily tested by isolating viruses from apparently recombinant plaques, and retesting them for a stable recombinant phenotype by plaque assay.

4.2 Genetic Recombination Between Viral RNA Genomes

Early genetic analysis revealed the existence of RNA recombination in picornaviruses (LEDINKO and HIRST 1961; COOPER 1969, 1977), and subsequent biochemical studies showed that recombinant viruses could be detected in mixed infections such that inherited proteins were derived from both parents ROMANOVA et al. 1980). However, it was not until 1982 that RNA recombination between viral genomes was conclusively demonstrated by fingerprinting of foot-and-mouth disease virus recombinants (KING et al. 1982).

KING (1987) has speculated that RNA recombination is important either to create genetic diversity and therefore evolutionary advantages for RNA viruses, or to eliminate deleterious mutations present in the genome due to the high error rate of RNA polymerase. In nature, recombination events produce faster-growing neurovirulent revertants of attenuated vaccine strains (KEW and NOTTAY 1984).

4.2.1 Characteristics of Poliovirus RNA Recombination

Three main characteristics of picornavirus RNA recombination have been elucidated. (a) Recombination is very efficient. KING (1987) has estimated that as much as 10% to 20% of viral RNA molecules undergo recombinational events in a single infectious cycle. KIRKEGAARD and BALTIMORE (1986) have measured recombination frequencies of 10^{-3} over a region of only 200 nucleotides. (b) RNA recombination is homologous. Viable recombinants do not harbor insertions, deletions, or base substitutions, even when the site of crossover contains heterologous sequences (KIRKEGAARD and BALTIMORE 1986). Furthermore, homologous recombination occurs at such a high frequency that similar rates of nonhomologous recombination, resulting in nonviable virus, would be incompatible with viral proliferation. (c) RNA recombination is not site specific. Recombination can occur at many different sequences in the viral genome and no structural elements in the RNA seem to be required (KIRKEGAARD and BALTIMORE 1986; KING 1987).

4.2.2 Mechanism of Poliovirus Recombination

Two mechanisms have been invoked to explain poliovirus RNA recombination. RNA rearrangements could occur by breaking and rejoining of RNA molecules, a mechanism used to splice cellular pre-mRNA molecules; alternatively, the viral RNA polymerase could switch to a second RNA template during RNA replication.

With the availability of defined genetic markers, KIRKEGAARD and BALTIMORE (1986) demonstrated that recombination occurs during negative-strand synthesis by a template switching of RNA polymerase, a copy choice mechanism. These experiments were performed by crossing guanidine-sensitive wild-type virus (g^s) with a variant of 3NC202 (g^r) (see Sect. 3.1.2) in a single cycle infection. When wild-type virus replication was allowed at the nonpermissive temperature for the 3NC202 variant, the two viruses recombined. However, when the replication of the 3NC202-variant but not of the wild-type virus was allowed (permissive temperature, in the presence of guanidine), no recombinants were observed. Because the replicating parental genome donated its 3' end, RNA recombination occurs during replication and the viral polymerase must switch templates during negative-strand synthesis.

It will be very interesting to develop an in vitro RNA recombination system which should allow the detailed analysis of viral and possible cellular components involved in RNA recombination (J. PATA and K. KIRKEGAARD, personal communication). Furthermore, such studies might reveal whether cellular RNA molecules undergo genetic recombination, although this has never been observed.

5 Genetic Complementation of Poliovirus

Complementation describes the interaction between viral gene products in cells infected with two mutants, resulting in the enhanced yield of one (one-way complementation) or both (two-way complementation) mutants while their genotypes remain unchanged. Classically, mutants which cannot complement each other have been considered to have defects in the same gene product. The division of mutants into complementation groups has given rough estimates of the number of genes involved in a particular genetic process.

5.1 The Complementation Test

Complementation tests are performed by infecting cells either singly or mixedly with two mutants which display distinguishable phenotypes under nonpermissive conditions. The individual virus yields from these single cycle infections are then determined by plaque assay. A complementation index for a mutant is calculated as the true yield of the mutant (corrected for reversion) in a mixed infection divided by the yield of the mutant in a single infection. Very often, spontaneous revertants represent a significant fraction of the total virus yield obtained in infections and it is, therefore, important to correct for this value.

5.2 Genetic Complementation Among Poliovirus Mutants

Complementation has been observed among poliovirus mutants bearing lesions in the capsid region of the genome (COOPER 1969, 1977). It was found that defective interfering (DI) particles (COLE and BALTIMORE 1973) containing large deletions in the capsid regions of their genomes can be propagated in the presence of a helper virus. Furthermore, phenotypic mixing of structural proteins in cells mixedly infected with different poliovirus serotypes (LEDINKO and HIRST 1961) has supported the idea that structural proteins can be provided in trans. However, significant complementation among mutants bearing lesions in the nonstructural region had not been reported in the past (COOPER 1969, 1977). It has become clear that this was most likely due to the presence of multiple lesions in the parental mutant genome (see Sect. 3).

For well-defined mutants bearing lesions in 3A, 2A, and 2C (Sect. 3) and in one instance 3D (mutant ts035, AGUT et al. 1989), efficient complementation by other mutants has been demonstrated. On the other hand, for certain mutations in 2B, 3D (mutant 3D-3, BERNSTEIN et al. 1986) and the 3' noncoding region complementation has not been shown. Several explanations could account for the lack of complementation among the latter well-defined mutants. First, mutations in RNA sequence and structure are expected to be *cis* dominant. This is clearly the case with 3' noncoding region mutants (BERNSTEIN et al. 1986). Secondly, because the genome is translated as a polyprotein, single mutations could have pleiotropic effects. Structural alterations in the polyprotein could affect posttranslational processing, or a particular lesion could affect additional functions of precursor polypeptides. Thirdly, because some events in the viral life cycle occur in specialized cellular compartments (for example, replication complexes are attached to smooth membranes), there might not be free diffusion of certain viral proteins. Lastly, a viral polypeptide might associate immediately with the RNA from which it has been translated.

These results show that the nonstructural region of the poliovirus genome contains several individual genetic units. However, many more well-defined mutants will be necessary to establish a complete complementation map of the viral genome.

6 Concluding Remarks

The determination of the entire sequence of the poliovirus RNA genome, the observation that a cDNA copy of the viral genome is infectious, and the elucidation of the three-dimensional structure of the poliovirion have allowed the application of modern genetics to the study of this animal RNA virus. Many well-defined mutations have been introduced into the viral genome and mutant phenotypes have been studied. Structures in the RNA molecule and functions of individual genetic elements are under intense scrutiny to elucidate the molecular mechanism of viral translation, replication and neurovirulence. In addition, mutants have been isolated that fail to inhibit host cell translation, suggesting that defined mutants can also be used as a tool to study the macromolecular pathways in the mammalian host cells.

Acknowledgements. We wish to thank Karla Kirkegaard for communication of results prior to publication and for many helpful comments on the manuscript. We are also grateful to Kyle Johnson, Dennis Macejak, and Theresa Najita for critical reading of the manuscript, and especially to Erika Norris for the artwork.

References

Agut H, Matzukura T, Bellocq C, Dreano M, Nicolas JC, Girard M (1981) Isolation and preliminary characterization of temperature-sensitive mutants of poliovirus type 1. Ann Inst Pasteur 132E: 445–460

Agut H, Kean K, Fichot O, Morasco J, Flanegan JB, Girard M (1989) A point mutation in the poliovirus polymerase gene determines a complementable temperature-sensitive defect of RNA replication. Virology 168: 302–311

Alexander HE, Koch G, Mountain DM, Sprunt K, Van Damme O (1958) Infectivity of ribonucleic acid of poliovirus on HeLa cell monolayers. Virology 5: 172–173

Anderson-Sillman K, Bartal S, Tershak DR (1984) Guanidine-resistent poliovirus mutants produce modified 37-kilodalton proteins. J Virol 50: 922–928

Argos P, Kamer G, Nicklin M, Wimmer E (1984) Similarity in gene organization and homololgy between proteins of animal picornaviruses and a plant comovirus suggest common ancestry of these virus families. Nucleic Acids Res 12: 7251–7267

Armstrong JA, Edmonds M, Nakazato H, Phillips BS, Vaughan MH (1972) Polyadenylic acid sequences in the virion RNA of poliovirus and eastern equine encephalitis virus. Science 176: 526–528

Arnold E, Luo M, Vriend G, Rossmann MG, Palmenberg AC, Parks GD, Nicklin MJH, Wimmer E (1987) Implications of the picornavirus capsid structure for polyprotein processing. Proc Natl Acad Sci USA 84: 21–25

Baltimore D (1968) Structure of the poliovirus replication intermediate RNA. J Mol Biol 32: 359–368

Bernstein HD, Baltimore D (1988) Poliovirus mutant that contains a cold-sensitive defect in viral RNA synthesis. J Virol 62: 2922–2928

Bernstein HD, Sonenberg N, Baltimore D (1985) Poliovirus mutant that does not selectively inhibit host cell protein synthesis. Mol Cell Biol 5: 2913–2923

Bernstein HD, Sarnow P, Baltimore D (1986) Genetic complementation among poliovirus mutants derived from an infectious cDNA clone. J Virol 60: 1040–1049

Bienz K, Egger D, Rasser Y, Bossart W (1983) Intracellular distribution of poliovirus proteins and the induction of virus-specific cytoplasmic structures. Virology 131: 39–48

Bienz K, Egger D, Pasamontes L (1987) Association of poliovirus proteins of the P2 genomic region with the viral replication complex and virus-induced membrane synthesis as visualized by electron microscopic immunocytochemistry and autoradiography. Virology 160: 220–226

Bienz K, Egger D, Troxler M, Pasamontes L (1990) Structural organization of poliovirus RNA replication is mediated by viral proteins of the P2 genomic region. J Virol 64: 1156–1163

Butterworth BE, Shimshick E, Yin FH (1976) Association of polioviral RNA polymerase complex with phospholipid membranes. J Virol 19: 457–466

Caliguiri IA, Tamm I (1968) Action of guanidine on the replication of poliovirus RNA. Virology 35: 408–417

Cech TR, Bass BL (1986) Biological catalysis by RNA. Ann Rev Biochem 55: 599–629.

Colbere-Garapin F, Christodoulou C, Crainic R, Garapin A-C, Candera A (1988) Addition of foreign oligopeptides to the major capsid protein of poliovirus. Proc Natl Acad Sci USA 85: 8668–8672

Cole CN, Baltimore D (1973) Defective interfering particles of poliovirus. II. Nature of the defect. J Mol Biol 76: 325–343

Cooper PD (1969) The genetic analysis of poliovirus. In: Levy BD (ed) Biochemistry of viruses. Dekker, New York, pp 177–218

Cooper PD (1977) Genetics of picornaviruses. In: Fraenkel-Conrat H, Wagner RR (eds) Comprehensive virology, vol 9. Plenum, New York, pp 133–207

Currey KM, Peterlin BM, Maizel JV Jr (1986) Secondary structure of poliovirus RNA: correlation of computer-predicted with electron microscopically observed structure. Virology 148: 33–46

Dahlberg AE, Dingman CW, Peacock AC (1969) Electrophoretic characterization of bacterial polyribosomes in agarose-acrylamide composite gels. J Mol Biol 41: 139–147

Dalbadie-McFarland G, Cohen LW, Riggs AD, Morin C, Itakura K, Richards JH (1982) Oligonucleotide-directed mutagenesis as a general and powerful method for studies of protein function. Proc Natl Acad Sci USA 79: 6409–6413

Davanloo P, Rosenberg AH, Dunn JJ, Studier PW (1984) Cloning and expression of the gene for bacteriophage T7 RNA polymerase. Proc Natl Acad Sci USA 81: 2035–2039

Dewalt PG, Semler BL (1987) Site-directed mutagenesis of proteinase 3C results in a poliovirus deficient in synthesis of viral RNA polymerase. J Virol 61: 2162–2170

Dildine SL, Semler BL (1989) The deletion of 41 proximal nucleotides reverts a poliovirus mutant containing a temperature-sensitive lesion in the 5′ noncoding region of genomic RNA. J Virol 63: 847–862

Dulbecco R, Vogt P (1954) Plaque formation and the isolation of pure lines with poliomyelitis virus. J Exp Med 99: 167–182

Emini EA, Leibowitz J, Diamond DC, Bonin J, Wimmer E (1984) Recombinants of Mahoney and Sabin strain poliovirus type 1: analysis of in vitro phenotypic markers and evidence that resistance to guanidine maps in the nonstructural proteins. Virology 137: 74–85

Etchison D, Milburn SC, Edery I, Sonenberg N, Hershey JWB (1982) Inhibition of HeLa cell protein synthesis following poliovirus infection correlates with the proteolysis of a 220,000-dalton polypeptide associated with eukaryotic initiation factor 3 and cap binding protein complex. J Biol Chem 257: 14806–14810

Etchison D, Hansen J, Ehrenfeld E, Edery I, Sonenberg N, Milburn S, Hershey JWB (1984) Demonstration in vitro that eucaryotic initiation factor 3 is active but that a cap-binding protein complex is incactive in poliovirus-infected HeLa cells. J Virol 51: 832–837

Evans DMA, Dunn G, Minor PD, Schild GC, Cann AJ, Stanway G, Almond JW, Currey K, Maizel JV (1985) A single nucleotide change in the 5′ non-coding region of the genome of the Sabin type 3 poliovaccine is associated with increased neurovirulence. Nature 314: 548–550

Hanecak R, Semler BL, Anderson CW, Wimmer E (1982) Proteolytic processing of poliovirus polypeptides: antibodies to polypeptide P3-7c inhibit cleavage at glutamine-glycine pairs. Proc Natl Acad Sci USA 79: 3973–3977

Hjerten S, Jerstedt S, Tiselius A (1965) Electrophoretic "particle sieving" in polyacrylamide gels as applied to ribosomes. Anal Biochem 11: 211–218

Hogle JM, Chow M, Filman DJ (1985) Three-dimensional structure of poliovirus at 2.9 Å resolution. Science 229: 1358–1365

Huang AS, Baltimore D (1970) Initiation of polysome formation in poliovirus-infected HeLa cells. J Mol Biol 47: 275–291

Iizuka N, Kuge S, Nomoto A (1987) Complete nucleotide sequence of the genome of coxsackievirus B1. Virology 156: 64–73

Johnson VH, Semler BL (1988) Defined recombinants of poliovirus and coxsackievirus: sequence-specific deletions and functional substitutions in the 5′-noncoding regions of viral RNAs. Virology 162: 47–57

Kalderon D, Oostra BA, Ely BK, Smith AE (1982) Deletion loop mutagenesis: a novel method for the construction of point mutations using deletion mutants. Nucleic Acids Res 10: 5161–5171

Kawamura N, Kohara M, Abe S, Komatsu T, Tago K, Arita M, Nomoto A (1989) Determinants in the 5′ noncoding region of poliovirus Sabin 1 RNA that influence the attenuation phenotype. J Virol 63: 1302–1309

Kew OM, Nottary BK (1984) Evolution of the oral polio vaccine strains in humans occur by both mutation and intramolecular recombination. In: Chanock R, Lerner R (eds) Modern approaches to vaccines. Cold Spring Harbor Laboratory, pp 357–362

King AMQ (1987) RNA Viruses do it. TIG 3: 60–61

King AMQ, McCahon D, Slade WR, Newman JWI (1982) Recombination in RNA. Cell 29: 921–928

Kirkegaard K (1990b) Mutations in VP1 of poliovirus specifically affect both encapsidation and release of viral RNA. J Virol 64: 195–206

Kirkegaard K, Baltimore D (1986) The mechanism of RNA recombination in poliovirus. Cell 47: 433–443

Kirkegaard K, Nelsen B (1990a) Conditional poliovirus mutants made by random deletion mutagenesis of infectious cDNA. J Virol 64: 185–194

Kitamura N, Semler BL, Rothberg PG, Larsen GR, Adler CJ, Dorner AJ, Emini EA, Hanecak R, Lee JJ, van der Werf S, Anderson CW, Wimmer E (1981) Primary structure, gene organization and polypeptide expression of poliovirus RNA. Nature 291: 547–533

Koenig H, Rosenwirth B (1988) Purification and characterization of poliovirus protease 2A by means of a functional assay. J Virol 62: 1243–1250

Konarska MM, Sharp PA (1986) Electrophoretic separation of complexes involved in the splicing of precursor to mRNAs. Cell 46: 845–855

Kräusslich HG, Nicklin MJH, Toyoda H, Etchison D, Wimmer E (1987) Poliovirus proteinase 2A induces cleavage of eucaryotic initiation factor 4F polypeptide p 220. J Virol 61: 2711–2718

Kuge S, Nomoto A (1987) Construction of viable deletion and insertion mutants of the Sabin strain of type 1 poliovirus: functional of the 5' noncoding sequence in viral replication. J Virol 61: 1478–1487

Kuge S, Kawamura N, Nomoto A (1989) Genetic variation occurring on the genome of an in vitro insertion mutant of poliovirus type 1. J Virol 63: 1069–1075

Kuhn RJ, Tada H, Ypma-Wong MF, Dunn JJ, Semler BL, Wimmer E (1988a) Construction of a "mutagenesis cartridge" for poliovirus genome-linked viral protein: isolation and characterization of viable and nonviable mutants. Proc Natl Acad Sci USA 85: 519–523

Kuhn RJ, Tada H, Ypma-Wong MF, Semler BL, Wimmer E (1988b) Mutational analysis of the genome-linked protein VPg of poliovirus. J Virol 62: 4207–4215

La Monica N, Almond JW, Racaniello VR (1987) A mouse model for poliovirus neurovirulence identifies mutations that attenuate the virus for humans. J Virol 61: 2917–2920

Larsen G, Semler BL, Wimmer E (1981) Stable hairpin structure within the 5'-terminal 85 nucleotides of poliovirus RNA. J Virol 37: 328–335

Ledinko N, Hirst GK (1961) Mixed infection of HeLa cells with polioviruses type 1 and 2. Virology 14: 207–219

Lee C-K, Wimmer E (1988) Proteolytic processing of poliovirus proteins: elimination of 2A^{pro}-mediated, alternative of polypeptide 3CD by in vitro mutagenesis. Virology 166: 405–416

Li J-P, Baltimore D (1988) Isolation of poliovirus 2C mutants defective in viral RNA synthesis. J Virol 62: 4016–4021

Li J-P, Baltimore D (1990) An intragenic revertant of a poliovirus 2C mutant has an uncoating effect. J Virol 64: 1102–1107

Lloyd RE, Grubman MJ, Ehrenfeld E (1988) Relationship of p 220 cleavage during picornavirus infection to 2A proteinase sequencing. J Virol 62: 4216–4223

Maniatis T, Fritsch EF, Sambrook J (1982) Molecular cloning: a laboratory manual. Cold Spring Harbor Laboratory, Cold Spring Harbor

McAllister WT, Morris C, Rosenberg AH, Studier FW (1981) Utilization of bacteriophage T7 late promoters in recombinant plasmids during infection. J Mol Biol 153: 527–544

Minor PD, Dunn G (1988) The effect of sequence in the 5' non-coding region on the replication of polioviruses in the human gut. J Gen Virol 69: 1091–1096

Nicklin MJH, Krausslich HHG, Toyoda H, Dunn JJ, Wimmer E (1987) Poliovirus polypeptide precursors: expression in vitro and processing by exogenous 3C and 2A proteinase. Proc Natl Acad Sci USA 84: 4002–4006

Noble J, Levintow L (1970) Dynamics of poliovirus specific RNA synthesis and the effects of inhibitors of virus replication. Virology 40: 634–642

Nomoto A, Detjen B, Pozzatti R, Wimmer E (1977a) The location of the polio genome protein in viral RNA and implication for RNA synthesis. Nature 268: 208–213

Nomoto A, Kitamura N, Golini F, Wimmer E (1977b) The 5'-terminal structures of poliovirion RNA and poliovirus mRNA differ only in the genome-linked protein VPg. Proc Natl Acad Sci USA 74: 5345–5349

Nomoto A, Omata T, Toyoda H, Kuge S, Horie H, Kataoka Y, Genba Y, Nakano Y, Imura N (1982) Complete nucleotide sequence of the attenuated poliovirus Sabin 1 strain genome. Proc Natl Acad Sci USA 79: 5793–5797

Omata Y, Kohara M, Sakai Y, Kameda A, Imura N, Nomoto A (1984) Cloned infectious complementary DNA of the poliovirus Sabin 1 genome: biochemical and biological properties of the recovered virus. Gene 32: 1–10

Omata Y, Kohara M, Abe S, Itoh H, Komatsu T, Arita M, Semler BL, Wimmer E, Kuge S, Makeda A, Nomoto A (1986) Construction of recombinant viruses between Mahoney and Sabin strains of type 1 poliovirus and their biological characteristics. In: Lerner RA, Chanock RM, Brown F (eds) Vaccines '85. Molecular and chemical basis of resistance to parasitic, bacterial, and viral diseases. Cold Spring Harbor Loboratory, New York, pp 279–283

Peden KWC, Nathans D (1982) Local mutagenesis within deletion loops of DNA heteroduplexes. Proc Natl Acad Sci USA 79: 7214–7217

Pelletier J, Sonenberg N (1988) Internal initiation of translation of eukaryotic mRNA directed by a sequence derived from poliovirus RNA. Nature 334: 320–325

Pelletier J, Kaplan G, Racaniello VR, Sonenberg N (1988) Cap-independent translation of poliovirus mRNA is conferred by sequence elements within the 5' noncoding region Mol Cell Biol 8: 1103–1112

Pettersson RF, Ambros V, Baltimore D (1978) Identification of a protein linked to nascent poliovirus RNA and to the polyuridylic acid of negative-strand RNA. J Virol 27: 357–365

Pilipenko EV, Blinov VM, Romanova LI, Sinyakov AN, Maslova SV, Agol VI (1989) Conserved structural domains in the 5'-untranslated region of picornaviral genomes: an analysis of the segment controlling translation and neurovirulence. Virology 168: 201–209

Pincus SE, Wimmer E (1986) Production of guanidine-resistant and -dependent poliovirus mutants from cloned cDNA: mutations in polypeptide 2C are directly responsible for altered guanidine sensitivity. J Virol 60: 793–796

Pincus SE, Diamond DC, Emini EA, Wimmer E (1986) Guanidine-selected mutants of poliovirus: mapping of point mutations to polypeptide 2C. J Virol 57: 638–646

Pincus SE, Kuhn RJ, Yang C-F, Toyoda H, Takeda N, Wimmer E (1987a) The poliovirus genome: a unique RNA in structure, gene orgaization, and replication. In: Inoye M, Dudock BS (eds) Molecular biology of RNA: new perspectives. Academic, San Diego, pp 175–210

Pincus SE, Rohl H, Wimmer E (1987b) Guanidine-dependent mutants of poliovirus: identification of three classes with different growth requirements. Virology 157: 83–88

Pleij CWA, Abrahams JP, van Belkum E, Rietveld K, Bosch L (1987) The spatial folding of the 3' noncoding region of aminoacylatable plant viral RNAs. In: Brinton MA, Rueckert PR (eds) Positive strand viruses. Liss, New York, pp 299–316

Putnak JR, Phillips BA (1981) Picornaviral structure and assembly. Microbiol Rev 45: 287–315

Racaniello VR, Baltimore D (1981a) Molecular cloning of poliovirus cDNA and determination of the complete nucleotide sequence of the viral genome. Proc Natl Acad Sci USA 78: 4887–4891

Racaniello VR, Baltimore D (1981b) Cloned poliovirus complementary DNA is infectious in mammalian cells. Science 214: 916–919

Racaniello VR, Meriam C (1986) Poliovirus temperature-sensitive mutant containing a single nucleotide deletion in the 5' -noncoding region of the viral RNA. Virology 155: 498–507

Rivera VM, Welsh JD, Maizel JV Jr (1988) Comparative sequence analysis of the 5' noncoding region of the enteroviruses and rhinoviruses. Virology 165: 42–50

Romanova LI, Tolskaya EA, Koleenikova MS, Agol VI (1980) Biochemical evidence for intertypic genetic recombination of polioviruses. FEBS Lett 118: 109–112

Rueckert RR (1985) Picornaviruses and their replication. In: Fields BN (ed) Virology. Raven, New York, pp 705–738

Rueckert RR, Wimmer E (1984) Systematic nomenclature of picornavirus proteins. J Virol 50: 957–959

Sarnow P (1989) Role of 3'-end sequences in infectivity of poliovirus transcripts made in vitro. J Virol 63: 467–470

Sarnow P, Bernstein HD, Baltimore D (1986) A poliovirus temperature-sensitive RNA synthesis mutant located in a noncoding region of the genome. Proc Natl Acad Sci USA 83: 571–575

Sedivy JM, Capone JP, Raj Bhandary UL, Sharp PA (1987) An inducible mammalian amber suppressor: propagation of a poliovirus mutant. Cell 50: 379–389

Semler BL, Anderson CW, Hanecak R, Dorner LF, Wimmer E (1982) A membrane-associated precursor to poliovirus VPg identified by immunoprecipitation with antibodies directed against a synthetic heptapeptide. Cell 28: 405–412

Semler BL, Dorner AJ, Wimmer E (1984) Production of infectious poliovirus from cloned cDNA is dramatically increased by SV40 transcription and replication signals. Nucleic Acids Res 12: 5123–5141

Semler BL, Johnson VH, Tracy S (1986) A chimeric plasmid from cDNA clones of poliovirus and coxsackievirus produces a recombinant virus that is temperature-sensitive. Proc Natl Acad Sci USA 83: 1777–1781

Shortle D, DiMaio D, Nathans D (1981) Directed mutagenesis. Annu Rev Genet 15: 265–294

Skinner MA, Racaniello VR, Dunn G, Cooper J, Minor P, Almond JW (1989) New model for the secondary structure of the 5' non-coding RNA of poliovirus is supported by biochemical and genetic data that also show that RNA secondary structure is important in neurovirulence. J Mol Biol 207: 379–392

Spector DH, Baltimore D (1974) Requirement of 3'-terminal poly(adenylic acid) for the infectivity of poliovirus RNA. Proc Natl Acad Sci USA 71: 2983–2987

Spector DH, Baltimore D (1975a) Polyadenylic acid on poliovirus RNA II. Poly(A) on intracellular RNA. J Virol 15: 1418–1431

Spector DH, Baltimore D (1975b) Polyadenylic acid on poliovirus RNA III. in vitro addition of polyadenylic acid to poliovirus RNAs. J Virol 15: 1432–1439

Stanway G, Cann AJ, Hauptmann R, Hughes P, Clarke LD, Mountford RC, Minor PD, Schild GC, Almond JW (1983) The nucleotide sequence of poliovirus type 3 Leon 12a₁b: comparison with poliovirus type 1. Nucleic Acids Res 11: 5629–5641

Stanway G, Hughes PJ, Mountford RC, Reeve P, Minor PD, Schild GC, Almond JW (1984) Comparison of the complete nucleotide sequences of the genomes of the neurovirulent polioviruses P3/Leon/37 and its attenuated Sabin vaccine derivative P3/Leon 12a$_1$b. Proc Natl Acad Sci USA 81: 1539–1543

Steinhauer DA, Holland JJ (1987) Rapid evolution of RNA viruses. Annu Rev Microbiol 41: 409–433

Svitkin YV, Maslova SV, Agol VI (1985) The genomes of attenuated and virulent poliovirus strains differ in their in vitro translation efficiencies. Virology 147: 243–252

Svitkin YV, Pestova, TV, Maslova SV, Agol VI (1988) Point mutations modify the response of poliovirus RNA to a translation initiation factor: a comparison of neurovirulent and attenuated strains. Virology 166: 394–404

Takegami T, Kuhn RJ, Anderson CW, Wimmer E (1983a) Membrane-dependent uridylylation of the genome-linked protein VPg of poliovirus. Proc Natl Acad Sci USA 80: 7447: 7451

Takegami T, Semler BL, Anderson CW, Wimmer E (1983b) Membrane fractions active in poliovirus RNA replication contain VPg precursor polypeptides. Virology 128: 33–47

Tershak DR (1982) Inhibition of poliovirus polymerase by guanidine in vitro. J Virol 41: 313–318

Tobin GJ, Young DC, Flanegan JB (1989) Self-catalyzed linkage of poliovirus terminal protein VPg to poliovirus RNA. Cell 59: 511–519

Toyoda H, Kohara M, Kataoka Y, Suganuma T, Omata T, Imura N, Nomoto A (1984) Complete nucleotide sequences of all three poliovirus serotype genomes: implication for genetic relationship gene function and antigenic determinants. J Mol Biol 174: 561–585

Toyoda H, Nicklin NJH, Murray MG, Anderson CW, Dunn JJ, Studier FW, Wimmer E (1986) A second-encoded proteinase involved in proteolytic processing of poliovirus polyprotein. Cell 45: 761–770

Trono D, Andino R, Baltimore D (1988a) An RNA sequence of hundreds of nucleotides at the 5′ end of poliovirus RNA is involved in allowing viral protein synthesis. J Virol 62: 2291–2299

Trono D, Pelletier J, Sonenberg N, Baltimore D (1988b) Translation in mammalian cells of a gene linked to the poliovirus 5′ noncoding region. Science 241: 445–448

van der Werf S, Bradley J, Wimmer E, Studier FW, Dunn JJ (1986) Synthesis of infectious poliovirus RNA by purified T7 RNA polymerase. Proc Natl Acad Sci USA 83: 2330–2334

van Dyke Ta, Flanegan JB (1980) Identification of poliovirus polypeptide p63 as a soluble RNA-dependent RNA polymerase. J Virol 35: 732–740

Weiner AM, Maizels N (1987) tRNA-like structures tag the 3′ ends of genomic RNA molecules for replication: implications for the origin of protein synthesis. Proc Natl Acad Sci USA 84: 7383–7387

Wimmer E, Kuhn RJ, Pincus S, Yang C-F, Toyoda H, Nicklin M, Takeda N (1987) Molecular events leading to picornavirus genome replication. In: Woolhouse HV, Ellis THN, Chater KF, Davies JW, Hull R (eds) Virus replication and genome interactions. Company of Biologists, Cambridge, pp 251–276

Wood WI, Gitschier J, Lasky IA, Lewis RM (1985) Base composition-independent hybridization in tetramethylammonium chloride: a method for oligonucleotide screening of highly complex gene libraries. Proc Natl Acad Sci USA 82: 1585–1588

Yogo Y, Wimmer E (1972) Polyadenylic acid at the 3′-terminus of poliovirus RNA. Plroc Natl Acad Sci USA 69: 1877–1882

Ypma-Wong MF, Dewalt PG, Johnson VH, Lamb, JG, Semler BL (1988a) Protein 3CD is the major poliovirus proteinase responsible for cleavage of the P1 capsid precursor. Virology 166: 265–270

Ypma-Wong MF, Filman DJ, Hogle JM, Semler BL (1988b) Structural domains of the poliovirus polyprotein are major determinants for proteolytic cleavage at gln-gly pairs. J Biol Chem 263: 17846–17856

Subject Index

Current Topics in Microbiology and Immunology

Volumes published since 1983 (and still available)

Vol. 138: **Goebel, Werner (Ed.):** Intracellular Bacteria. 1988. 18 figs. IX, 179 pp. ISBN 3-540-50001-4

Vol. 139: **Clarke, Adrienne E.; Wilson, Ian A. (Ed.):** Carbohydrate-Protein Interaction. 1988. 35 figs. IX, 152 pp. ISBN 3-540-19378-2

Vol. 140: **Podack, Eckhard R. (Ed.):** Cytotoxic Effector Mechanisms. 1989. 24 figs. VIII, 126 pp. ISBN 3-540-50057-X

Vol. 141: **Potter, Michael; Melchers, Fritz (Ed.):** Mechanisms in B-Cell Neoplasia 1988. Workshop at the National Cancer Institute, National Institutes of Health, Bethesda, MD, USA, March 23–25, 1988. 1988. 122 figs. XIV, 340 pp. ISBN 3-540-50212-2

Vol. 142: **Schüpbach, Jörg:** Human Retrovirology. Facts and Concepts. 1989. 24 figs. 115 pp. ISBN 3-540-50455-9

Vol. 143: **Haase, Ashley T.; Oldstone, Michael B. A. (Ed.):** In Situ Hybridization. 1989. 33 figs. XII, 90 pp. ISBN 3-540-50761-2

Vol. 144: **Knippers, Rolf; Levine, A. J. (Ed.):** Transforming Proteins of DNA Tumor Viruses. 1989. 85 figs. XIV, 300 pp. ISBN 3-540-50909-7

Vol. 145: **Oldstone, Michael B. A. (Ed.):** Molecular Mimicry. Cross-Reactivity between Microbes and Host Proteins as a Cause of Autoimmunity. 1989. 28 figs. VII, 141 pp. ISBN 3-540-50929-1

Vol. 146: **Mestecky, Jiri; McGhee, Jerry (Ed.):** New Strategies for Oral Immunization. International Symposium at the University of Alabama at Birmingham and Molecular Engineering Associates, Inc. Birmingham, AL, USA, March 21–22, 1988. 1989. 22 figs. IX, 237 pp. ISBN 3-540-50841-4

Vol. 147: **Vogt, Peter K. (Ed.):** Oncogenes. Selected Reviews. 1989. 8 figs. VII, 172 pp. ISBN 3-540-51050-8

Vol. 148: **Vogt, Peter K. (Ed.):** Oncogenes and Retroviruses. Selected Reviews. 1989. XII, 134 pp. ISBN 3-540-51051-6

Vol. 149: **Shen-Ong, Grace L. C.; Potter, Michael; Copeland, Neal G. (Ed.):** Mechanisms in Myeloid Tumorigenesis. Workshop at the National Cancer Institute, National Institutes of Health, Bethesda, MD, USA, March 22, 1988. 1989. 42 figs. X, 172 pp. ISBN 3-540-50968-2

Vol. 150: **Jann, Klaus; Jann, Barbara (Ed.):** Bacterial Capsules. 1989. 33 figs. XII, 176 pp. ISBN 3-540-51049-4

Vol. 151: **Jann, Klaus; Jann, Barbara (Ed.):** Bacterial Adhesins. 1990. 23 figs. XII, 192 pp. ISBN 3-540-51052-4

Vol. 152: **Bosma, Melvin J.; Phillips, Robert A.; Schuler, Walter (Ed.):** The Scid Mouse. Characterization and Potential Uses. EMBO Workshop held at the Basel Institute for Immunology, Basel, Switzerland, February 20–22, 1989. 1989. 72 figs. XII, 263 pp. ISBN 3-540-51512-7

Vol. 153: **Lambris, John D. (Ed.):** The Third Component of Complement. Chemistry and Biology. 1989. 38 figs. X, 251 pp. ISBN 3-540-51513-5

Vol. 154: **McDougall, James K. (Ed.):** Cytomegaloviruses. 1990. 58 figs. IX, 286 pp. ISBN 3-540-51514-3

Vol. 155: **Kaufmann, Stefan H. E. (Ed.):** T-Cell Paradigms in Parasitic and Bacterial Infections. 1990. 24 figs. IX, 162 pp. ISBN 3-540-51515-1

Vol. 156: **Dyrberg, Thomas (Ed.):** The Role of Viruses and the Immune System in Diabetes Mellitus. 1990. 15 figs. XI, 142 pp. ISBN 3-540-51918-1

Vol. 157: **Swanstrom, Ronald; Vogt, Peter K. (Ed.):** Retroviruses. Strategies of Replication. 1990. 40 figs. XII, 260 pp. ISBN 3-540-51895-9

Vol. 158: **Muzyczka, Nicholas (Ed.):** Viral Expression Vectors. 1990. approx. 20 figs. approx. XII, 190 pp. ISBN 3-540-52431-2

Vol. 159: **Gray, David; Sprent, Jonathan (Ed.):** Immunological Memory. 1990. 38 figs. XII, 156 pp. ISBN 3-540-51921-1

Vol. 160: **Oldstone, Michael B.A.; Koprowski, Hilary (Ed.):** Retrovirus Infections of the Nervous System. 1990. 16 figs. XII, 176 pp. ISBN 3-540-51939-4